Marcel Schumm

ZnO-based semiconductors studied by Raman spectroscopy

Marcel Schumm

ZnO-based semiconductors studied by Raman spectroscopy

Semimagnetic alloying, doping and nanostructures

Südwestdeutscher Verlag für Hochschulschriften

Impressum/Imprint (nur für Deutschland/ only for Germany)

Bibliografische Information der Deutschen Nationalbibliothek: Die Deutsche Nationalbibliothek verzeichnet diese Publikation in der Deutschen Nationalbibliografie; detaillierte bibliografische Daten sind im Internet über http://dnb.d-nb.de abrufbar.
Alle in diesem Buch genannten Marken und Produktnamen unterliegen warenzeichen-, marken- oder patentrechtlichem Schutz bzw. sind Warenzeichen oder eingetragene Warenzeichen der jeweiligen Inhaber. Die Wiedergabe von Marken, Produktnamen, Gebrauchsnamen, Handelsnamen, Warenbezeichnungen u.s.w. in diesem Werk berechtigt auch ohne besondere Kennzeichnung nicht zu der Annahme, dass solche Namen im Sinne der Warenzeichen- und Markenschutzgesetzgebung als frei zu betrachten wären und daher von jedermann benutzt werden dürften.

Verlag: Südwestdeutscher Verlag für Hochschulschriften Aktiengesellschaft & Co. KG
Dudweiler Landstr. 99, 66123 Saarbrücken, Deutschland
Telefon +49 681 37 20 271-1, Telefax +49 681 37 20 271-0, Email: info@svh-verlag.de
Zugl.: Würzburg, Julius Maximilians Universität, Diss., 2009

Herstellung in Deutschland:
Schaltungsdienst Lange o.H.G., Berlin
Books on Demand GmbH, Norderstedt
Reha GmbH, Saarbrücken
Amazon Distribution GmbH, Leipzig
ISBN: 978-3-8381-1009-7

Imprint (only for USA, GB)

Bibliographic information published by the Deutsche Nationalbibliothek: The Deutsche Nationalbibliothek lists this publication in the Deutsche Nationalbibliografie; detailed bibliographic data are available in the Internet at http://dnb.d-nb.de.
Any brand names and product names mentioned in this book are subject to trademark, brand or patent protection and are trademarks or registered trademarks of their respective holders. The use of brand names, product names, common names, trade names, product descriptions etc. even without a particular marking in this works is in no way to be construed to mean that such names may be regarded as unrestricted in respect of trademark and brand protection legislation and could thus be used by anyone.

Publisher:
Südwestdeutscher Verlag für Hochschulschriften Aktiengesellschaft & Co. KG
Dudweiler Landstr. 99, 66123 Saarbrücken, Germany
Phone +49 681 37 20 271-1, Fax +49 681 37 20 271-0, Email: info@svh-verlag.de

Printed in the U.S.A.
Printed in the U.K. by (see last page)
ISBN: 978-3-8381-1009-7

Contents

List of Figures

List of Tables

Chapter 1

Introduction

In the past years, the scientific interest in ZnO was renewed because improved processing and better theoretical understanding raised hope for various new applications, for instance in optoelectronics, spintronics, and nanotechnology [Jagadish 2006, Klingshirn 2007, Ozgur 2005]. ZnO is a II-VI semiconductor compound with a large, direct band gap of 3.4 eV. Its large exciton binding energy of 60 meV makes it very attractive for use in optoelectronic devices such as light-emitting diodes or laser diodes emitting in the blue and ultraviolet spectral range. P-type doping of ZnO is a major issue, which is strongly related to these optoelectronic applications. Among the potential impurities acting as acceptors, the group-V elements nitrogen, phosphorus, arsenic, and antimony are the most promising [Look 2006]. The successful p-type doping of ZnO via nitrogen incorporation and the subsequent fabrication of blue ZnO-based light emitting diodes was achieved using temperature-modulated molecular beam epitaxy [Tsukazaki 2004]. Still, no straightforward procedure has been established yet for fabricating reproducible and stable p-type ZnO of high quality.

Theoretical calculations predicted p-type $Zn_{1-x}Mn_xO$ as well as n-type ZnO alloyed with other transition metal ions as candidates for future spintronics applications with stable ferromagnetic configurations above room temperature [Dietl 2000, Sato 2001]. In addition, a model was proposed for ferromagnetic coupling in n-type, magnetically diluted oxides such as ZnO, SnO_2, and TiO_2 due to bound magnetic polarons [Coey 2005]. However, the experimental situation regarding the magnetic properties of transition-metal-alloyed ZnO is ambiguous and often even contradicting [Norton 2006].

Much of the future potential of ZnO lies in the field of nanotechnology with applications as nanolasers, nanosensors, and nano field-effect-transistors [Wang 2004]. Because ZnO has a strong tendency to self-organized growth, nanostructures of various different morphologies like nanoparticles, nanowires, and nanobelts can be obtained by straightforward fabrication.

The manipulation of ZnO material properties by the incorporation of impurities (e.g. doping, magnetic alloying) or by miniaturization (e.g. nanostructures) strongly affects the crystal structure of ZnO and therewith its vibrational properties. To study the modified lattice dynamics, Raman spectroscopy is an excellent, non-invasive technique, which is applied as major research method for this thesis. In most experiments, the elementary excitations detected by Raman scattering are phonons. In that case, it delivers information on structural properties such as chemical composition, orientation, or crystalline quality. Additionally, electronic and magnetic properties can be addressed, e.g. by Raman resonance effects and Raman scattering from magnons, respectively. The Raman scattering results of this thesis are complemented by other experimental methods, e.g. X-ray diffraction, photoluminescence, electron paramagnetic resonance, and transmission electron microscopy.

This thesis comprises two parts. Part I, 'Basics', starts with chapter 2, in which the fundamentals of the Raman scattering theory as well as the mainly used Raman setups are presented. The general material properties and important applications of ZnO are outlined in chapter 3, with special focus on the lattice dynamics and Raman scattering characteristics.

Part II of this thesis, 'Results and discussion', begins with chapter 4, which is devoted to the study of pure ZnO by Raman spectroscopic means. An important question addressed is to which extent the crystal quality of the samples is affected by different morphologies or different growth processes. Bulk ZnO single crystals of high structural quality are characterized, which are also used as host crystals for implanted ZnO systems. In addition, disorder effects are analyzed on microcrystalline ZnO and Ar-irradiated ZnO single crystals. Finally, ZnO nanoparticles of very small average diameter are investigated with regard to their structural properties, organic ligands, and size effects.

In chapter 5, ZnO alloyed with transition metal elements is studied by Raman spectroscopy and complementary methods. The corresponding samples were fabricated with varying transition metal concentrations, using several different growth methods such as molecular beam epitaxy, vapor phase transport, and ion implantation. A key question for such diluted magnetic semiconductors is whether the material contains uniformly distributed transition metal ions on the appropriate atom site or if secondary phases are responsible for the observed magnetic properties. Therefore, the experiments analyze the influence of several parameters on the structural quality of the samples and on the tendency to form precipitates of secondary phases. Among these parameters are the transition metal species (vanadium, manganese, iron, cobalt, and nickel), the transition metal concentration (0.2 at.% - 32 at.%), and the annealing temperature (100 °C - 900 °C in air or vacuum).

The challenge of ZnO p-doping is addressed in chapter 6, which focuses on the impact of nitrogen dopants on the ZnO lattice dynamics. In the literature, several additional modes are reported for the Raman spectra of nitrogen-doped ZnO, e.g. [Wang 2001], but their origin is still ambiguous and heavily contested. The main question is whether the additional features are localized vibrations of substitutional nitrogen on oxygen sites or if they correspond to disorder-induced Raman scattering. Using Raman spectroscopy and complementary methods, the additional modes as well as the structural impact of nitrogen incorporation on the ZnO crystal are studied for ZnO:N samples fabricated by ion implantation as well as epitaxial growth.

Finally, the major results of this thesis are summarized in the chapters 7 and 8 in English and German, respectively.

Part I

Basics

Chapter 2

Raman spectroscopy

Raman spectroscopy is a commonly used optical and mostly non-invasive research method, for example for chemical analysis or in solid state physics. Amongst others, it has been successfully applied to investigate semiconductor systems [Cardona 2007], e.g. in heterostructures and at interfaces [Esser 1996, Geurts 1996]. Raman spectroscopy is based upon the inelastic scattering of monochromatic light within the studied sample, accompanied by the generation or annihilation of elementary excitations. In most experiments for this thesis, the elementary excitations are vibrations, particularly lattice vibrations (phonons). Raman spectroscopy then provides access to the lattice dynamics of a sample and therewith delivers information on structural properties like chemical composition, orientation, or crystalline quality. In addition, electronic and magnetic properties can be addressed, e.g. by Raman resonance effects and Raman scattering from magnons, respectively.

All this information is gathered by the analysis of the Raman signals with regard to their frequency position, frequency width, recorded intensity, and line shape in the Raman spectra. In a typical spectrum, the intensity is plotted versus the so-called Raman shift. The former is proportional to the number of photons of a certain frequency reaching the detector, whereas the Raman shift is given by the frequency difference between the scattered light and the monochromatic excitation source. This shift corresponds to the energy of the generated or annihilated elementary excitation. It is usually given in wavenumber $\bar{\nu} = \nu/c$, where $[\bar{\nu}] = \mathrm{cm}^{-1}$ ($1\ \mathrm{cm}^{-1} \cong$ 300 GHz $\cong 0.124$ meV). Conventionally, if the inelastic light scattering generates an elementary excitation and the scattered light therefore exhibits a lower frequency, the Raman shift is denoted positive (Stokes Raman scattering). In the case of an annihilated elementary excitation, the wavenumber value becomes negative (anti-Stokes Raman scattering). In experiments with T $\leq$ room temperature, usually only a few phonons are thermally excited and so the Stokes process

Energy

excited electronic state

virtual electronic states

$\hbar\omega_i$ $\hbar\omega_i$ $\hbar\omega_i$ $\hbar\omega_i^*$

$\hbar\Omega_s$

excited phonon state Ω_1
phonon ground state $\Omega = 0$

	Anti-Stokes Raman scattering:	Rayleigh scattering:	Stokes Raman scattering:	Resonant Stokes Raman scattering:
Energy of scattered light:	$\hbar\omega_i + \hbar\Omega_s$	$\hbar\omega_i$	$\hbar\omega_i - \hbar\Omega_s$	$\hbar\omega_i^* - \hbar\Omega_s$
Raman shift:	$-\hbar\Omega_s$	0	$+\hbar\Omega_s$	$+\hbar\Omega_s$

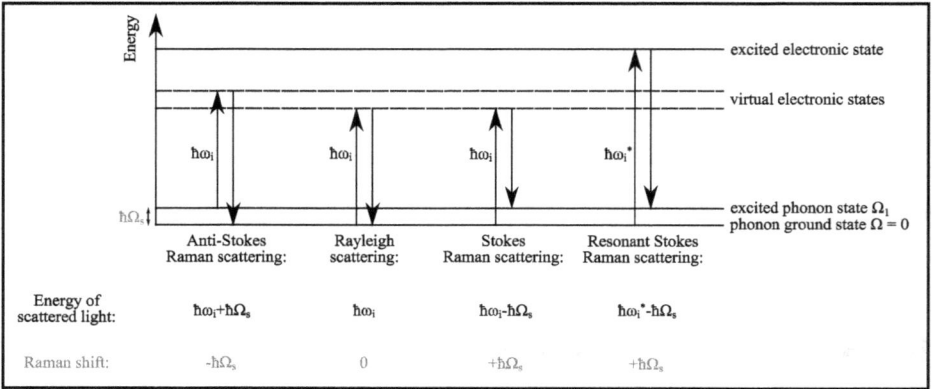

Figure 2.1: *Energy diagram for inelastic (Raman) and elastic (Rayleigh) scattering. While in the case of resonant Raman scattering an actual electronic transition is involved, the other Raman processes are described by the introduction of virtual electronic states. The frequency difference between the scattered light and the monochromatic excitation source is conventionally called Raman shift. Therefore, a Raman shift of zero corresponds to elastic Rayleigh scattering.*

is dominant. In Figure 2.1, the inelastic Raman scattering processes (Stokes and anti-Stokes) and the elastic Rayleigh scattering are illustrated in an energy diagram using the concept of virtual electronic states. Also shown is the resonant Raman scattering process, in which the energy of the incident light coincides with a natural electronic transition of the studied system, causing the Raman scattering probability to increase significantly (see subsection 2.1.2).

Summing up, the x-axis value of a signal in a Raman spectrum corresponds to the energy of the generated (or annihilated) elementary excitation and the y-axis value corresponds to the detected intensity of scattered light with this energy (see Figure 2.2). In this work, only the part of the spectrum with the Stokes Raman signals is shown (positive Raman shift values). Additionally, the spectra are recorded with a small offset to the Rayleigh scattering peak at 0 cm^{-1}. With an intensity share of about 99.9%, the Rayleigh signal could saturate the detector and swamp the much smaller Raman signals.

The following section 2.1 outlines the basic theory of Raman scattering. The experimental setups used for the Raman measurements in this thesis as well as the micro- and macro-Raman scattering techniques are treated in section 2.2. Application of Raman spectroscopy in the characterization of semiconductor systems in general is presented in section 2.3, while Raman scattering on ZnO is described in detail in chapter 3.

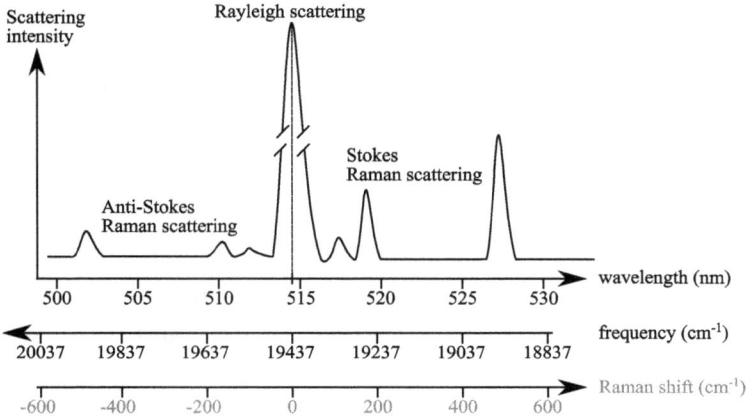

Figure 2.2: *Schematic Raman spectrum with a green (514.5 nm) laser as excitation source. The intensity value is proportional to the numbers of detected photons of a certain frequency. Raman shift is the frequency difference between the inelastically scattered light and the monochromatic excitation. As indicated, the elastic Rayleigh scattering (with about 99.9% intensity share) is by far the dominant process.*

2.1 Raman scattering fundamentals

This section is devoted to the theory of Raman scattering, covering the basic principles like interaction mechanisms for inelastic light scattering (2.1.1), resonance aspects (2.1.2), and symmetry-imposed restrictions (2.1.3), based on the concepts presented in detailed treatises on Raman scattering [Brüesch 1986, Cardona 2007, Esser 1996, Geurts 1996, Hayes 1979, Richter 1976, Yu 1999].

2.1.1 Principles of Raman scattering theory

Raman scattering as an inelastic scattering process is characterized by an energy transfer between the incident photon of energy $\hbar\omega_i$ and the sample via the generation or annihilation of elementary excitations, which are assumed to be phonons within this theoretical treatment. The energy of the scattered photon is $\hbar\omega_s \neq \hbar\omega_i$. Energy conservation demands that the energy transfer equals the energy of the generated or annihilated elementary excitation $\hbar\Omega_s$:

Stokes process (phonon generation):

$$\hbar\omega_i - \hbar\omega_s = \hbar\Omega_s. \tag{2.1}$$

Anti-Stokes process (phonon annihilation):

$$\hbar\omega_s - \hbar\omega_i = \hbar\Omega_s. \tag{2.2}$$

Generally, the frequency of the incident light is much higher (about 100x) than the frequency of the scattered excitation. Thus, the frequencies of the incident and the scattered light differ only in the low percentage range:

$$\omega_i \gg \Omega_s \Rightarrow \omega_i \approx \omega_s. \tag{2.3}$$

The wave vectors of the incident light $(\vec{k}_i)$, of the scattered light $(\vec{k}_s)$, and of the elementary excitation $(\vec{q}_j)$ are correlated by the conservation of the quasi-momentum:

Stokes process (phonon generation):

$$\vec{k}_i - \vec{k}_s = \vec{q}_j. \tag{2.4}$$

Anti-Stokes process (phonon annihilation):

$$\vec{k}_s - \vec{k}_i = \vec{q}_j. \tag{2.5}$$

As a result of energy and quasi-momentum conservation, the elementary excitations involved in Raman scattering are characterized by well-defined (Ω, q) pairs. In 180° backscattering geometry (as applied in all experiments for this thesis), the wave vectors can be simplified using the scalar form. For Stokes scattering and with c as the velocity of light and n(ω) as the index of refraction, this leads to:

$$q_j = \frac{1}{c}(n(\omega_i)\omega_i - n(\omega_s)\omega_s). \tag{2.6}$$

Generally, the wavelength λ_i of the incident (and also λ_s of the scattered) light is much longer than the lattice constant a_0 of the crystal. Therefore, k_i and k_s, and due to equations 2.4 and

2.5 also q_j, are much smaller than the wave vector of the Brillouin zone boundary $2\pi/a_0$. In first order phonon Raman scattering, the wave vector of the phonons is then given by

$$q_j \approx 0. \tag{2.7}$$

For one-phonon Raman scattering, this implies that the phonon wave vector lies within the inner few percents of the Brillouin zone. In multi-phonon scattering, only the sum of the phonon wave vectors must be close to zero. Thus, phonons from outside the Brillouin zone center can be involved. Examples for such multi-phonon scattering in the case of ZnO are described in section 3.1.2. Limitations of the conservation of quasi-momentum and therefore of equation 2.7 can be caused by crystal imperfections (see subsection 2.1.3).

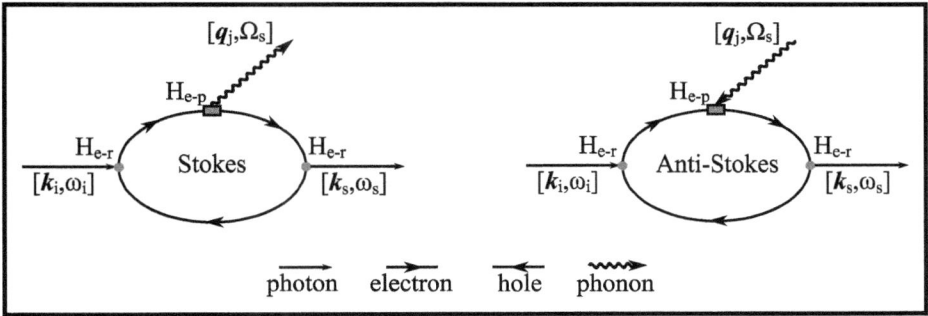

Figure 2.3: *Feynman diagrams of first order Stokes and anti-Stokes Raman scattering. The interaction between the incident light and the lattice is mediated by the generation of an electron-hole pair. H_{e-r} is the Hamiltonian of the electron-radiation interaction, H_{e-p} of the electron-phonon interaction. Note that the displayed diagrams represent only one of six scattering processes which contribute to one-phonon Stokes and anti-Stokes Raman scattering, respectively [Yu 1999].*

The interaction between the incident light and phonons in Raman scattering is not direct, but mediated by electronic interband transitions. This can be illustrated by Feynman diagrams of first order Stokes and anti-Stokes processes, see Figure 2.3. In the microscopic picture, the interaction can be described by the following processes:

(a) Absorption of the incident photon $\hbar\omega_i$ with excitation of the initial electronic state $|0\rangle$ into an intermediate electronic state (electron-hole pair) $|n\rangle$.

(b) Electron-phonon interaction: Scattering of the electron-hole pair into a new state $|n'\rangle$ with emission or absorption of a phonon $\hbar\Omega_j$.

(c) Recombination of the electron-hole pair into the initial electronic state $|0\rangle$ with emission of the scattered photon $\hbar\omega_s$.

The initial electronic state $|0\rangle$ is also the final electronic state, hence, the electrons remain unchanged by the Raman scattering process. Using third-order perturbation theory [Yu 1999], the scattering probability corresponding to Figure 2.3 is approximated by:

$$P_{sc}(\omega_i, \omega_s) = \frac{2\pi}{\hbar} \left| \sum_{n,n'} \frac{\langle 0|H_{e-r}(\omega_s)|n'\rangle \, \langle n'|H_{e-p}|n\rangle \, \langle n|H_{e-r}(\omega_i)|0\rangle}{(\hbar\omega_i - (E_n - E_0))(\hbar\omega_s - (E_{n'} - E_0))} \right|^2 \delta[\hbar\omega_i - \hbar\omega_0 - \hbar\omega_s], \quad (2.8)$$

where H_{e-r} and H_{e-p} are Hamiltonians corresponding to the electron-radiation and the electron-phonon interaction, respectively. $|0\rangle$ denotes the initial, $|n\rangle$ and $|n'\rangle$ intermediate electronic states, with the corresponding energies E_0, E_n, and $E_{n'}$. E_n and $E_{n'}$ represent virtual electronic energy levels, thus, no energy conservation is required for the electronic transitions $(\hbar\omega_i - (E_n - E_0))$ and $(\hbar\omega_s - (E_{n'} - E_0))$ in the denominator of equation 2.8.

2.1.2   Resonant Raman scattering

Although the excited electronic states in general are virtual states, the scattering probability in equation 2.8 diverges if actual electronic levels with energy E_n or $E_{n'}$ exist. The corresponding Raman signals show a strongly enhanced intensity if the number of such electronic states is large (e.g. at critical points in the band structure). If it is the incident light, which coincides with a natural electronic transition, the resonance is called incoming resonance. Accordingly, outgoing resonance is achieved if the scattered light has the appropriate energy.

Resonant Raman spectroscopy possesses several benefits, for example:

(i) By resonance enhancement, Raman spectroscopy of surfaces and interfaces can be possible with monolayer sensitivity [Cardona 2007, Esser 1996, Muck 2004, Wagner 2002].

(ii) For the study of diluted nanostructures, resonant Raman spectroscopy can be combined with micro-Raman spectroscopy (2.2.2), providing the required detection sensitivity and the ability to determine the lateral distribution of the nanosized structures.

(iii) Material selectivity in multilayered structures can be achieved by selection of the appropriate resonant excitation wavelengths.

(iv) The electronic properties of a sample are reflected by its resonance effect.

An example for these advantages is displayed in Figure 2.4: By a combination of micro-Raman scattering and resonant Raman spectroscopy, graphene monolayer flakes with an area of a few square micron are identified and distinguished from graphite layers. These experiments can be carried out in only a few minutes due to a double resonance caused by the special electronic properties of such small carbon layers [Ferrari 2006]. Several Raman resonance effects in ZnO based materials are discussed in the course of this thesis, especially in section 4.1 and section 5.2.

2.1.3 Selection rules and Raman tensor

For one-phonon Raman scattering, the mediated radiation-phonon interaction is only possible for optical phonons restricted to the center of the Brillouin zone (subsection 2.1.1). However, these are not the only restrictions for phonons in Raman scattering. Based on group theory, Raman selection rules can be derived, which determine whether phonons are Raman active, i.e. whether the corresponding Raman process has a non-zero scattering intensity. In the classical picture, the Raman scattering intensity i_R can be written as:

$$i_R \propto |\vec{\Pi}_i \, \Re \, \vec{\Pi}_s|^2, \tag{2.9}$$

with the polarizations $\vec{\Pi}_i$ and $\vec{\Pi}_s$ of the incident and scattered radiation, respectively, and the so-called Raman tensor $\Re$ [Yu 1999]. The latter describes the change of the susceptibility χ during a vibration with amplitude $\vec{A}$:

$$\Re = (\partial \chi / \partial \vec{A}) \vec{A}. \tag{2.10}$$

From 2.10 it follows that a phonon mode is Raman active if it induces a change of the

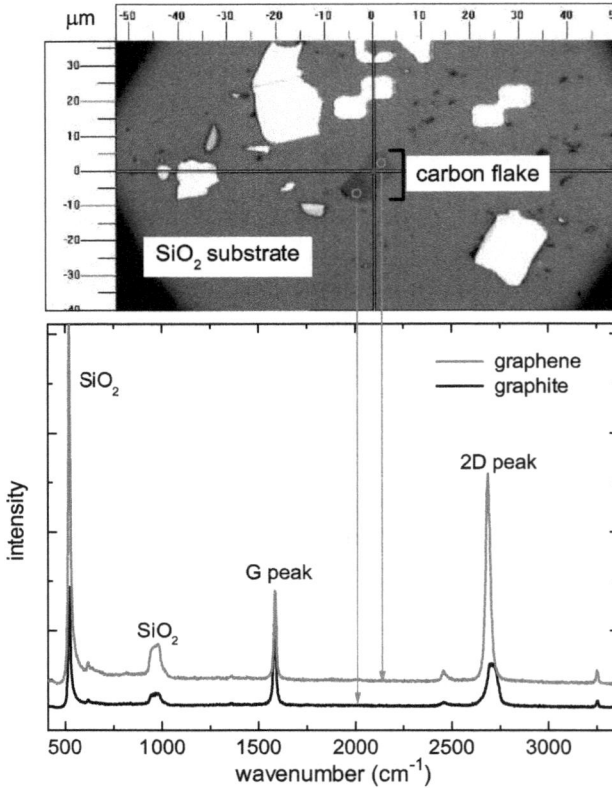

Figure 2.4: *Identification of a graphene flake on SiO_2 substrate by Raman scattering: Due to the special electronic properties of carbon layers with few monolayers thickness, a double resonance allows Raman scattering to distinguish graphene monolayer flakes from thicker graphite flakes by the peak ratio 2D/G and the FWHM of the 2D peak [Ferrari 2006] in experiments of only a few minutes duration. Note that also the signal of the underlying SiO_2 substrate is stronger when the focus lies on the (thinner) graphene.*

Raman polarizability $\partial\chi/\partial\vec{A}$. By contrast, a phonon mode is IR active (i.e. observable via infrared absorption) if it induces a change in the dipole moment [Brüesch 1986]. For systems with inversion center, IR absorption and Raman scattering are complementary methods, while for other systems there can be modes which are both IR and Raman active. Phonon modes which are neither Raman nor IR active are called silent modes. If one neglects the (generally small) frequency difference between the incident and the scattered light and if the system is non-magnetic, $\Re$ can be approximated by a symmetric tensor of rank two:

$$\Re = \begin{pmatrix} \Re_{xx} & \Re_{xy} & \Re_{xz} \\ \Re_{yx} & \Re_{yy} & \Re_{yz} \\ \Re_{zx} & \Re_{zy} & \Re_{zz} \end{pmatrix} \tag{2.11}$$

In the so called Porto notation [Damen 1966], the relevant information on the experimental configuration, i.e. polarization and direction of incident and scattered light relative to the sample orientation, is summarized by $\vec{k}_i(\vec{\Pi}_i, \vec{\Pi}_s)\vec{k}_s$. In combination with the structure and symmetry of the studied crystal, these vectors determine which Raman scattering processes are allowed in a given configuration.

If a crystal lattice has a center of inversion, each phonon mode can be classified as either odd, i.e. 180° phase shift of the phonon mode after an inversion, or even, i.e. phonon mode invariance against inversion. The coupling of the electronic states $|n\rangle$ and $|n'\rangle$ of equal symmetry with phonon generation or annihilation as described in subsection 2.1.1, equation 2.8, is only possible for even phonon modes. Consequently, the electronic mediation of the radiation-phonon interaction and therefore the Raman scattering process is not allowed for odd modes. In the rocksalt crystal structure, for example, all optical phonon modes show odd symmetry and therefore no one-phonon Raman scattering is allowed [Brüesch 1986], at least in perfect crystals. Crystal imperfections can lead to a softening of the Raman selection rules due to reduced symmetry. Examples of the resulting effects on Raman spectra can be found in subsection 2.3.2 and are discussed throughout the chapters 4, 5, and 6. The Raman selection rules for zinc blende and especially for the wurtzite structure of ZnO are presented in detail in subsection 3.1.2.

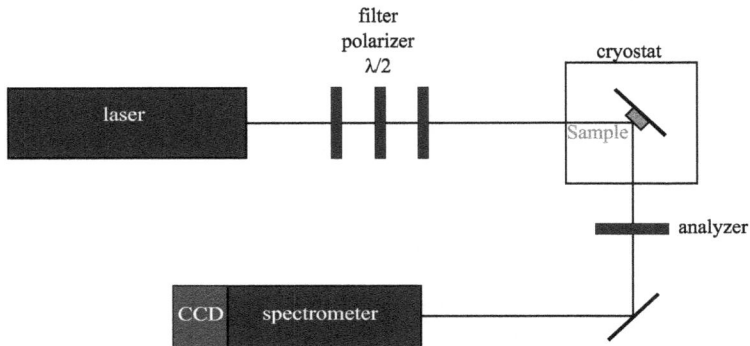

Figure 2.5: *Schematic experimental setup for Raman scattering experiments. Monochromatic laser light is focused on a sample, the scattered light is collected, and analyzed by a spectrometer and a detector, for example a CCD.*

2.2 Raman techniques and experimental setups

In subsection 2.2.1, the general setup of Raman experiments is introduced. As most of the Raman experiments presented in this work were conducted in micro-Raman scattering configuration, the advantages and disadvantages of such experiments compared to conventional macro-Raman scattering are discussed in subsection 2.2.2. The two setups mainly used for this work are described in detail in subsection 2.2.3.

More technical and theoretical information on laser spectroscopy and Raman scattering techniques can be found in [Demtröder 2002] and [McCreery 2000], respectively. Other experimental methods used (e.g. XRD and EPR) are described in detail when the respective results are discussed.

2.2.1 General setup of Raman experiments

The general setup for Raman scattering experiments is shown in Figure 2.5 and consists of (i) a monochromatic light source for excitation, usually a laser, (ii) optical equipment to bring the laser beam on the sample and collect the scattered light, (iii) a spectrometer to analyze the scattered light, and (iv) a detector to collect the signal. Complementary information on the sample may be derived if the exciting and analyzed light are manipulated with optical filters, polarizers, etc.

Today, light sources in Raman scattering setups are usually realized by laser systems. Gas lasers, solid-state lasers, dye lasers, and other laser devices provide a quasi-continuous variety of wavelengths from the IR to the UV as well as continuous wave (cw) power outputs from μW to several W. In this work, mainly the standard lines of cw gas lasers (argon ion, krypton ion, and helium-neon) were applied from the red to the UV spectral range. For magneto-Raman and low temperature Raman experiments, samples were placed in a sample chamber within a liquid helium bath or a liquid helium continuous flow cryostat. The choice of the spectrometer setup can result in a trade-off between sensitivity and resolution, as will be discussed in subsection 2.2.3. For the detection, multi-channel detectors are essential and for this thesis CCD arrays were used.

2.2.2 Micro- and macro-Raman scattering

The main difference between micro-Raman experiments and conventional macro-Raman is the insertion of an optical microscope in the experimental setup. Using a beam-splitter, the laser beam is injected into the collection axis of the optical microscope and focused on the sample using a microscope objective. This lens also collects the scattered light, which is then led back through the microscope, to the beam-splitter, and finally into the spectrometer. The corresponding setup is schematically shown in Figure 2.6.

The advantages of the micro-Raman technique include a lateral and to some extent also a depth resolution of the Raman scattering signal. Thus, it is possible to study the homogeneity of a sample surface and to discover small precipitates or other crystal defects in the micron range (corresponding to the laser spot size on the sample). In addition, different layers within a layered sample can be addressed using different focus depths, again in the micron range (depending on the confocal character of the microscope setup). The use of an optical microscope also gives an opportunity to detect and record inhomogeneities on the sample surface optically. As a further advantage, microscope lenses achieve higher collection efficiencies than a conventional macro-Raman apparatus.

However, this collection efficiency comes with the disadvantage of a less well-defined direction of the collected light. Hence, the backscattering conditions are lifted to a substantial degree. Another major disadvantage of micro-Raman compared to macro-Raman setups is the high power density in the laser spot because the size of the focused spot is usually in the μm range compared to mm for macro-Raman. This high power leads to local heating in the sample, which induces temperature effects, especially in low temperature measurements, and can even damage samples. For the material systems analyzed in this thesis, local heating was mostly uncritical. In

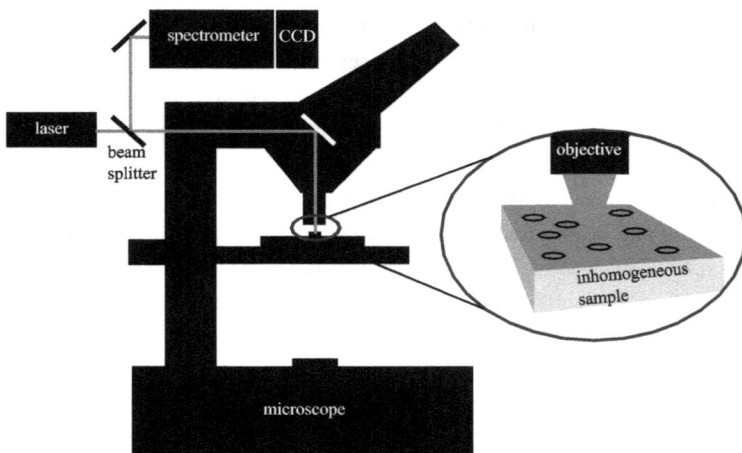

Figure 2.6: *Setup for micro-Raman scattering experiments. Using a beam-splitter, the exciting laser light is injected into an optical microscope and focused on the sample by an objective. The scattered light is collected, led to the beam-splitter, and into the spectrometer.*

the case of some temperature sensitive samples, however, the local heating effect was taken into account, especially in the case of wet-chemically synthesized nanoparticles with organic ligands (section 4.2 and subsection 5.2.3).

Further theoretical and technical details concerning the micro-Raman scattering technique are described in [McCreery 2000, Turrell 1996].

2.2.3   Setups: Dilor XY and Renishaw 1000

For part of the Raman measurements, the scattered light was analyzed by a triple monochromator (Dilor XY) in multi-channel mode and detected by a liquid-nitrogen-cooled CCD array. Besides its good spectral resolution (1 cm^{-1}/pixel in low dispersion and 0.3 cm^{-1}/pixel in high dispersion mode), the major advantage of this setup is its enormous versatility. Using an argon ion laser, not only its standard lines and multi-line UV output are provided, but also dye lasers can be pumped to deliver continuously tunable laser lines throughout the visible spectrum. Furthermore, the setup can be used in macro-Raman as well as micro-Raman configuration, both including a liquid helium cryostat. The helium bath cryostat (Oxford) of the macro-Raman configuration can also be utilized as magneto-cryostat delivering magnetic fields of up to 6 Tesla. The continuous-flow helium cryostat (CryoVac) of the micro-Raman setup is equipped with a temperature control

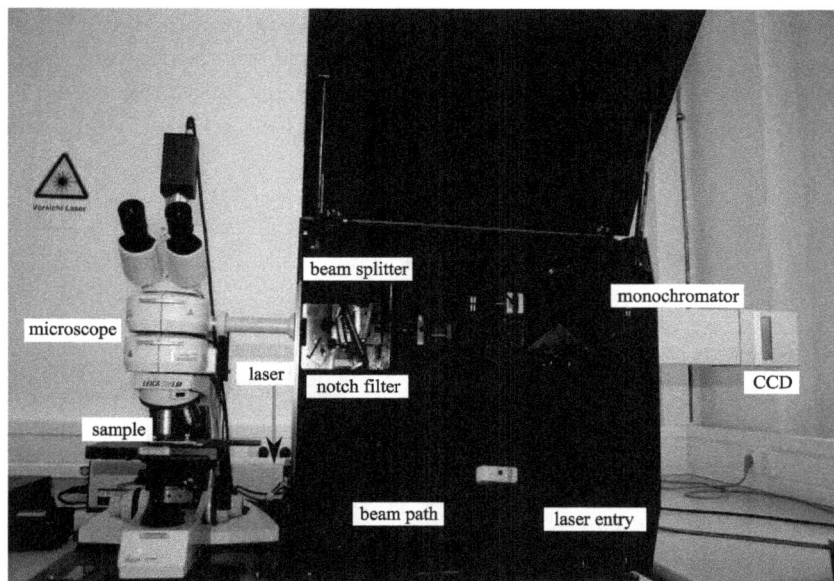

Figure 2.7: *Raman system Renishaw 1000 with Leica Microscope DM LM.*

system so that the sample temperature can be tuned continuously between about 10 K and room temperature. This setup includes an Olympus BHT microscope, equipped with several objectives (10x, 50x ULWD, 80x ULWD, and 100x). In the experiments involving the Dilor setup, mostly the standard lines (514.5 nm, 496.5 nm, 488.0 nm, 476.5 nm, 457.9 nm) of the argon ion laser were used as excitation source.

The other of the two mainly used setups is a Renishaw Raman system RM 1000 with a Leica DM LM microscope, which includes an integrated camera for optical photography. For this setup, excitation is limited to the 514.5 nm line of an argon ion laser and the 632.8 nm line of a helium-neon laser. To focus the laser beam, a 50x objective is used. The scattered light is analyzed by a single monochromator, equipped with a double notch filter system, and the signal is detected by a Peltier-cooled CCD array. Figure 2.7 shows this relatively small-sized table-top Raman system. The major advantage of this Renishaw system is its outstanding light throughput due to the use of only one monochromator. If its resolution of about 4 cm^{-1}/pixel is sufficient, this setup is capable of conducting Raman experiments much faster. Due to its high sensitivity, also temperature sensitive samples (e.g. nanoparticles with organic ligands) can be studied, because measurements at low laser powers (< 1 mW) still deliver good signal-to-noise ratios for most experiments.

The advantages and disadvantages of the Raman setups Renishaw RM 1000 and Dilor XY are summarized in Table 2.1. For fast experiments, lateral mapping, or for temperature sensitive samples, the Renishaw setup proved ideal. For more sophisticated experiments requiring high resolution, specific wavelengths, low temperature, or magnetic fields, the Dilor XY setup was used. Because of different depth-of-field values, the intensity ratios in the Raman spectra of layered samples taken with the two different setups can vary.

	Renishaw RM 1000	Dilor XY
Resolution	+	+++
Sensitivity	+++	+
Sample protection	+++	+
Excitation line variety	+	+++
Magneto-Raman	-	+++
Low temperature	-	+++
Optical photography	+++	+

Table 2.1: *Comparison of the advantages and disadvantages of the two mainly used Raman setups for this thesis. While the Dilor XY system is more versatile and has better spectral resolution, the high sensitivity of the Renishaw system allows fast experiments and high sample protection due to low laser power density.*

2.3 Raman spectroscopy on semiconductors

Raman scattering is a widely used technique for the characterization of semiconductor systems, whereas in most of the experiments, the studied elementary excitations are phonons. In the following subsections, the lattice dynamics of semiconductor crystals (2.3.1) and the application of Raman spectroscopy for the characterization of semiconductors (2.3.2) are outlined.

2.3.1 Raman scattering by lattice vibrations

Following [Yu 1999], the Hamiltonian describing the nuclear motion in a solid state lattice can be expressed as

$$H_{ion}(\vec{R}_1, ..., \vec{R}_n) = \sum_j \frac{P_j^2}{2M_j} + \sum_{j,j',j \neq j'} \frac{1}{2} \frac{Z_j Z_{j'} e^2}{|\vec{R}_j - \vec{R}_{j'}|} - \sum_{i,j} \frac{Z_j e^2}{|\vec{r}_i - \vec{R}_j|}, \qquad (2.12)$$

where $\vec{R}_j$ is the position, $\vec{P}_j$ the momentum, Z_j the charge, and M_j the mass of the nucleus j. In this model, the nuclei and core electrons (position $\vec{r}_j$) are combined to 'ions', which interact with the separated valence electrons. In the so-called Born-Oppenheimer approximation, the valence electrons follow the much heavier ions adiabatically and the 'ions' only see a time-averaged electronic potential of the valence electrons:

$$H_{ion} = \sum_j \frac{P_j^2}{2M_j} + E_e(\vec{R}_1, ..., \vec{R}_n). \qquad (2.13)$$

H_{ion} can be expanded as a function of the ion displacements, and the nuclei can be treated as an ensemble of harmonic oscillators. Then, the change of the ion Hamiltonian of equation 2.13 due to the displacement of ion k in unit cell l is given by:

$$H'_{\vec{u}_{kl}} = \frac{1}{2} M_k (\frac{d\vec{u}_{kl}}{dt})^2 + \frac{1}{2} \sum_{k'l'} \vec{u}_{kl} \Phi(kl, k'l') \vec{u}_{k'l'}. \qquad (2.14)$$

$\Phi(kl, k'l')$ is a second rank tensor comprising the force constants of the ion-ion interaction. In the quantum mechanical treatment, the vibrations in a crystal lattice are energetically quantized,

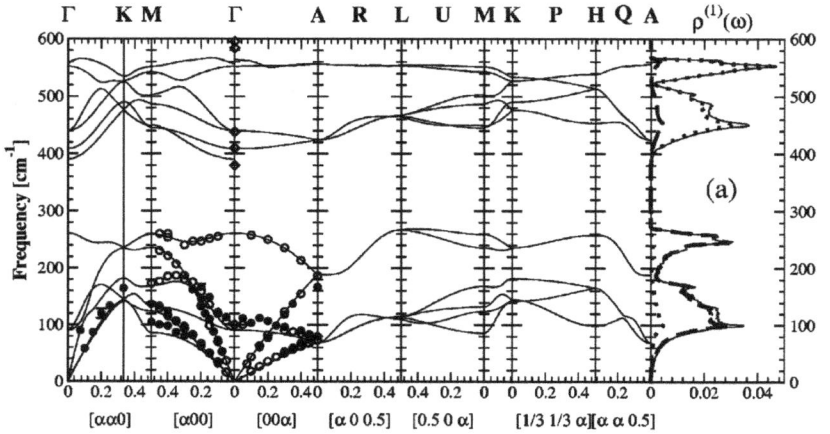

Figure 2.8: *Phonon dispersion relation for wurtzite ZnO from [Serrano 2004]. The energy (here: frequency) values of the wurtzite phonon modes are plotted versus their wavevector along high-symmetry directions of the crystal. Experimental data points by Raman scattering [Serrano 2003] and inelastic neutron scattering [Hewat 1970, Thoma 1974] are inserted as diamonds and circles, respectively. On the right hand, labeled (a), the one-phonon density of states is shown.*

the corresponding elementary excitations are called phonons. They can be described as plane waves:

$$\vec{u}_{kl}(\vec{q}, \omega) = \vec{u}_{k0} e^{(\vec{q}\vec{R}_l - \omega t)}. \tag{2.15}$$

The phonon is defined by its wave vector $\vec{q}$ and its frequency ω. Equation 2.15 reflects the translational symmetry: If two wave vectors are different by a whole-number multiple of the reciprocal lattice vector $\vec{R}_l$, they are physically equivalent.

The lattice dynamics of a semiconductor are reflected in its phonon dispersion relation (PDR), in which the energy of a lattice vibration is plotted versus its wave vector along high-symmetry directions of the crystal. Due to the translational symmetry in equation 2.15, the PDR is conventionally displayed within the Brillouin zone. While the PDR along high-symmetry directions of the crystal can be measured by inelastic neutron scattering for the entire Brillouin zone, Raman scattering can only give experimental data on the optical phonon energies at the center (see subsection 2.1.1). Several theoretical models to calculate PDR curves are presented in [Yu 1999]. As an example, the PDR of wurtzite ZnO from [Serrano 2004] is shown in Figure 2.8. More detailed information on the phonon modes displayed in the ZnO PDR is given in subsection 3.1.2.

If the number of states in the PDR is integrated over energy, the result is called phonon density of states (PDOS), see the one-phonon DOS curve (a) in Figure 2.8, for example. As discussed before, of all energy-wavenumber combinations given in the PDR, only optical phonons at the center of the Brillouin zone can participate in one-phonon Raman scattering in perfect crystals (subsection 2.1.1). Thus, the Raman shifts observed in the Raman spectra of this thesis mostly correspond to the energy of the optical phonons in the PRD curves at $q = 0$. In multi-phonon Raman scattering, though, only $\sum_i \vec{q_i} = 0$ is required and therefore also single phonons with $q \neq 0$ can participate in the overall process. Furthermore, if the crystal shows imperfections, a relaxation of the Raman selection rules can be observed due to reduced symmetry. This becomes important for the application of Raman spectroscopy in the characterization of semiconductor crystals, see subsection 2.3.2.

2.3.2 Raman analysis of semiconductor systems

A. Compound composition

Material identification of semiconductor crystals can be achieved by the analysis of Raman signals corresponding to the energy of the Raman-active optical phonon modes at the Brillouin zone center in the phonon dispersion relation (subsection 2.3.1). The PDR is well known for all common elemental and binary compound semiconductors. The identification becomes more difficult if the semiconductor crystal is composed of more than two different elements, for example a ternary compound $A_{1-x}B_xC$. In such compounds, atoms of the element B substitute atoms of the element A in the compound AC. The vibrational properties of such systems can be described by the modified random-element isodisplacement model (MREI), see [Anastassakis 1991, Chang 1968, Peterson 1986]. If the concentration of B on A sites is very low, the B atoms can be treated as isolated impurities (see paragraph B). For higher concentrations, three different mode behaviors can account for the observed vibrations of many ternary semiconductor compounds $A_{1-x}B_xC$ (see also Figure 2.9):

One-mode behavior:
With increasing concentration x of the element B, the modes of the binary compound AC show a continuous transition into the modes of the binary compound BC, and vice versa.
Example: $Cd_{1-x}Zn_xS$

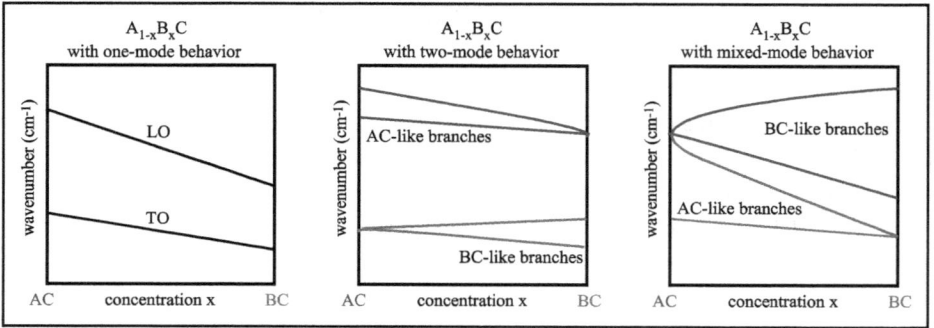

Figure 2.9: *One-mode, two-mode, and mixed-mode behavior of ternary $A_{1-x}B_xC$ semiconductor compounds.*

Two-mode behavior:

With increasing concentration x of the element B, the mode of the isolated impurity B in AC shows a transition into the longitudinal optical (LO) and transverse optical (TO) modes of BC, and vice versa.

Example: $Be_{1-x}Zn_xSe$

Mixed-mode behavior:

The modes show two-mode behavior and, additionally, a BC mode coincides with the impurity mode of A in BC, or vice versa.

Example: $Zn_{1-x}Mn_xTe$

The incorporation of a B atom into a binary compound AC can be regarded as a disturbance of the perfect AC crystal lattice. Therefore, besides these mode shifts in mixed crystals, also disorder effects can be expected as described in the next paragraph.

B. Crystal imperfections

Crystal defects can be classified as point defects, line defects, and complexes, depending on whether they involve single atoms, rows of atoms, or an ensemble of atoms, respectively. Typical

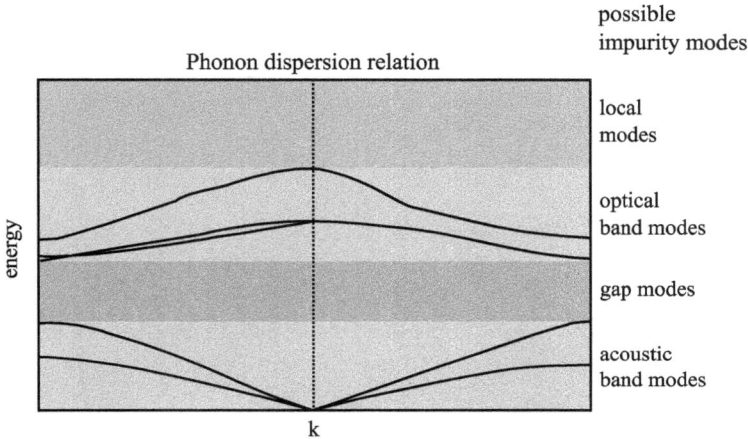

Figure 2.10: *Depending on the relation between the masses of the substitutional atom and the substituted host atom, impurity modes can occur as local modes above the optical phonon branches, as gap modes between the acoustic and the optical branches, or as band modes within the optical or acoustic wavenumber range.*

point defects in a crystal are vacancies, interstitial atoms, and substitutional atoms [Yu 1999]. While a vacancy is an intrinsic effect, interstitials and substitutions can be extrinsic effects, i.e. caused by a foreign atom. Such impurities may be incorporated on purpose to manipulate the magnetic properties (magnetic alloying) or the electronic properties (doping) of a semiconductor, for example ZnO alloyed with transition metal ions (chapter 5) or doped with nitrogen (chapter 6). Besides the intended effect, such impurities will also affect the crystal structure and the vibrational properties. To achieve the intended electronic or magnetic properties, the foreign atoms have to be incorporated substitutionally on the appropriate atom site and interstitials must be avoided. Such substitutional atoms may have their own vibrational signature. If a large amount of atoms is substituted, the system can be described in terms of a ternary compound with one-, two-, or mixed-mode behavior (subsection 2.3.2, A). In the case of low concentrations of B atoms in $A_{1-x}B_xC$, they can be regarded as isolated impurities. Depending on (i) the masses of atom B and the substituted atom A, (ii) the bond strength between the impurity and the host crystal, and (iii) the vibrational properties of the AC host lattice, such an impurity can show characteristic impurity modes.

Let ω_B be the vibrational frequency of the impurity B substituting A atoms in a binary AC host crystal with optical modes between $\omega_{opt,min}$ and $\omega_{opt,max}$ and acoustic modes between $\omega_{ac,min}$ and $\omega_{ac,max}$. Then, three types of impurity modes can be distinguished (see also Figure 2.10):

Local modes: $\omega_B > \omega_{opt,max}$

Gap modes: $\omega_{ac,max} < \omega_B < \omega_{opt,min}$

Band modes: $\omega_{opt,min} < \omega_B* < \omega_{opt,max}$ or $\omega_{ac,min} < \omega_B* < \omega_{ac,max}$

Both local modes and gap modes are localized at the position of the impurity atom. They can not couple to the vibrational spectra of the host lattice because no host vibrations with nearby frequencies exist. In the case of local modes, the frequency of the impurity is higher than the highest optical modes of the host, implying that the impurity is lighter than the substituted host atom: $m_B < m_A$. For $m_B > m_A$, an independent impurity mode is still possible if ω_B lies in the gap between the acoustic and the optical branch (provided such a gap exists!). For $m_B \approx m_A$, the impurity vibration lies within the frequency range of the optical or acoustic phonon branches of the host material and, therefore, the vibration couples to the adjacent host lattice phonons. In contrast to local and gap modes, the resulting band modes are collective lattice vibrations without localized character (but induced by a localized impurity). All these findings are relevant for the harmonic crystal. In the anharmonic crystal, additional impurity modes are possible [Sievers 1988].

If impurity modes fulfill the selection rules, Raman spectroscopy can give valuable information as to the substitutional character of the incorporated impurities in semiconductors [McCluskey 2000, Mayur 1996]. Vacancies, interstitials, and complexes have an effect on the vibrational properties of a crystal, too, whereas an accurate assignment is generally more difficult. While interstitials and complexes may also have characteristic vibrational signatures as in the case of substitutional impurity modes, all crystal defects reduce the crystal symmetry and therefore cause a softening of the Raman selection rules (subsection 2.1.3). The restriction to the Brillouin zone center (quasi-momentum conservation) is relaxed and thus, also phonons near the Brillouin zone center with $q \neq 0$ can participate in the inelastic scattering process. Depending on the dispersion of the phonon around $q = 0$, this will lead to peak broadening (often asymmetric) and peak shifting (towards lower wavenumber) of the Raman signals. In the spatial correlation model described in [Parayanthal 1984], the contribution of non-zero q-values to the Raman scattering intensity decays exponentially upon increasing wave vector with e^{-qL_c}, where q is the wave vector and L_c the correlation length. To induce a pronounced peak shifting and broadening, the crystal has to be strongly disturbed ($L_c \approx nm$). Even for bulk crystals with higher quality, however, crystal imperfections can still be observable: Phonon modes may occur in the Raman spectra which are symmetry-forbidden in an ideal crystal.

A similar effect is the optical phonon confinement in small crystals [Englman 1966, Ruppin 1970]. For optically isotropic materials, the confinement of the optical phonons in nanocrystals will lead to peak broadening and red shifting of the corresponding bulk phonon modes in the Raman spectra, comparable to a disturbed crystal with nm correlation length [Richter 1981]. For anisotropic materials, confined optical phonon modes different from the bulk modes can be expected [Fonoberov 2004]. Raman experiments on ZnO nanocrystals and the discussion of the observed peak shifting are presented in subsection 4.2.

C. Band structure properties

Besides structural information, also electronic properties can be derived from Raman scattering. Equation 2.8 in subsection 2.1.1 describes the Raman scattering intensity including the electron-phonon interaction H_{e-p}. For optical phonons, two major electron-lattice effects account for this interaction: Deformation potential scattering (LO and TO phonons) and the Fröhlich interaction (LO phonons in polar crystals) [Yu 1999].

Deformation potential:
The lattice deformation by phonons induces an additional time-modulated potential, i.e. the optical phonons alter the electronic energies by changing the bond lengths and bond angles.

Fröhlich interaction:
In polar or partly ionic crystals, a LO phonon implicates a uniform atom displacement within the unit cell. This relative displacement of opposite charges generates a macroscopic Coulomb potential. Its interaction with the electronic system is called Fröhlich interaction.

The character of these electron-phonon interactions can influence the Raman spectra. For example, due to Fröhlich scattering, impurities in doped semiconductors can be studied via the LO phonon interaction with the charged impurities [Esser 1996]. The influence of Fröhlich scattering on the Raman spectra of Mn-alloyed ZnO will be discussed in section 5.2.

As equation 2.8 in subsection 2.1.1 involves the summation over many electronic states, the deduction of electronic properties from Raman spectra can be arbitrarily complicated. It can be simplified if one or few of these states are dominant, i.e. if Raman resonance occurs. By varying the incident wavelength around the resonant energy (e.g. in vicinity of a band gap or of free or bound excitons), the change in the Raman efficiency reflects critical points of the electronic band structure. More examples of electronic properties observable by Raman scattering are discussed in [Esser 1996].

D. Magnetic Properties

Raman spectroscopy also provides access to the magnetic properties of diluted magnetic semiconductors (DMS), especially via spin-flip Raman spectroscopy [Petrou 1983, Ramdas 1982]. Also, Raman scattering from magnons was observed in magnetic systems. In the course of this work, no spin-flip Raman spectroscopy was applied, but scattering from magnons was observed (section 5.3). Therefore, only the theory of the latter will be shortly outlined.

Magnons are quantized excitations (spin waves) of a periodic spin system from its fully aligned ground state. While in phonon Raman scattering the interaction between the radiation and the lattice is mediated, a direct magnetic-dipole coupling was identified as possible process for one-magnon Raman scattering [Bass 1960]. However, by comparison to experimental data, another interaction was found to be the dominant process: electric-dipole coupling via spin-orbit interaction [Elliott 1963]. This indirect coupling due to mixing of spin and orbital motions fits to the experimentally observed intensities and symmetry restrictions. In contrast to the mostly symmetric phonon scattering, it is purely antisymmetric. The indirect electric-dipole coupling via spin-orbit interaction also gives rise to higher-order scattering, but only with intensities that are orders of magnitudes smaller. In two-magnon scattering of antiferromagnetic systems, yet another interaction is dominant: excited-state exchange interaction between opposite sublattices of the antiferromagnet [Fleury 1968], which can lead to an even stronger scattering intensity than the first-order effect. Theoretical details and experimental examples on both one- and two-magnon Raman scattering can be found in [Fleury 1968]. By observing magnons via Raman scattering, important magnetic properties of a system can be revealed, e.g. the temperature-dependent alignment of a spin system.

Chapter 3

Zinc oxide: Material properties and applications

The II-VI semiconductor compound ZnO has been studied for decades. Besides specific physical properties (e.g. large direct band gap of 3.4 eV and high exciton binding energy of 60 meV), some of the most obvious advantages of ZnO are its great availability, low cost production, and low toxicity. With its band gap corresponding to UV light of 365 nm, ZnO is transparent throughout the visible spectrum. Under normal conditions, it crystallizes in the hexagonal wurtzite structure. In conventional industrial applications, ZnO is used in large-scale manufacturing of cosmetics, plastics, or food additives, just to mention a few. In the past years, the scientific interest in ZnO was renewed because improved processing and better theoretical understanding raised hope for various new applications, for example in optoelectronics, spintronics, and nanotechnology. Despite enormous scientific effort in the recent years, important technical issues are yet to be solved, most prominently the p-type doping of ZnO. In the following sections 3.1 and 3.2, important material properties of ZnO as well as potential applications and processing issues related to the studied samples are described.

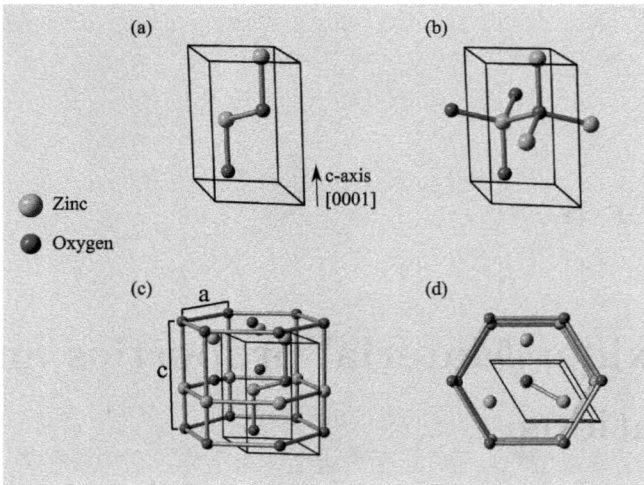

Figure 3.1: *ZnO with wurtzite crystal structure: (a) four atoms in the unit cell (two of each atom sort), (b) tetrahedral coordination, (c) hexagonal symmetry and the lattice parameters a and c, (d) top view of (c).*

3.1   Material properties

Here, only the material properties are presented, which are most important for this thesis. For more information, see some of the recent comprehensive reviews on ZnO material properties [Jagadish 2006, Klingshirn 2007, Ozgur 2005]. The crystal lattice and optical properties of ZnO are reviewed in the subsections 3.1.1 and 3.1.3, respectively. Lattice dynamics and Raman scattering on ZnO are discussed in detail in subsection 3.1.2.

3.1.1   Crystal structure and chemical binding

The chemical binding character of ZnO lies between covalent and ionic. Due to the large ionicity of the bonds between Zn and O atoms (about 0.62 on the Phillips scale), the two binding partners can be denoted as Zn^{2+} and O^{2-} ions, respectively [Chelikowsky 1986, Phillips 1970]. In ambient conditions, the thermodynamically stable phase of ZnO is the hexagonal wurtzite structure. Under stress or upon growth on cubic substrates, ZnO can also exhibit rock-salt structure [Bates 1962, Serrano 2004] or zinc-blende structure [Ashrafi 2000], respectively. However, all

ZnO systems studied for this thesis crystallized in wurtzite structure.

The wurtzite structure belongs to the space group P63mc. Figure 3.1 shows some of its characteristics: (a) It has four atoms in the unit cell, two of each atom sort. (b) Each atom is tetrahedrally coordinated, i.e. the four next neighbors are of the other atom sort and located at the edges of a tetrahedron. (c) The lattice has hexagonal symmetry and is characterized by the lattice parameters a and c. The two atom types occupy one hexagonal-close-packed sublattice each, displaced along the c-axis. The volume of the unit cell is approximately 47.6 Å^3. Consequently, ZnO has about 8.4 x10^{22} atoms/cm^{-3}.

While the ideal wurtzite structure shows a c/a-ratio of 1.63 [Ozgur 2005], the lattice constants of real wurtzite ZnO depend on impurities, external stress, temperature, etc. Pure, ordered ZnO at ambient conditions has lattice constants of a = 0.325 nm and c = 0.521 nm with a c/a-ratio of about 1.60 [Reeber 1970]. The orientation of a wurtzite sample is denoted by four-digit miller indices (h k i l), with h + k = -i. In this thesis, most samples with well-defined crystal orientation are c-plane oriented, i.e. the surface is the hexagonal (0001) plane. Correspondingly, the directions parallel to the c-axis are denoted with [0001], see also Figure 3.1.

3.1.2   Lattice vibrations and Raman scattering

In this section, the lattice dynamics of wurtzite ZnO are discussed with special focus on the Raman-active optical phonon modes. The wurtzite-type lattice structure of ZnO implies a basic unit of 4 atoms in the unit cell (2 Zn-O molecular units). Due to the number of n = 4 atoms in the unit cell, the number of phonons amounts to 3n = 12, with 3 acoustic modes (1xLA, 2xTA) and 3n - 3 = 9 optical phonons (3xLO, 6xTO). At the Γ-Point of the Brillouin zone, the optical phonons have the irreducible representation $\Gamma_{opt} = A_1 + 2B_1 + E_1 + 2E_2$ [Damen 1966], whereas the E modes are twofold degenerate. The B_1 modes are silent, i.e. IR and Raman inactive, and the E_2 branches are Raman active only. The Raman- and IR-active branches A_1 and E_1 are polar and therefore each splits into LO and TO modes with different frequencies due to the macroscopic electric fields of the LO phonons.

The lattice dynamics of a crystal are reflected in its phonon dispersion relation (subsection 2.3.1). Figure 3.2 shows the phonon dispersion relation of wurtzite ZnO for selected directions in the Brillouin zone [Serrano 2003]. The zone-center optical mode frequencies lie between about 100 cm^{-1} (12.5 meV) and 600 cm^{-1} (75 meV). Obviously, in the highest frequency range (550 cm^{-1} to 600 cm^{-1}), a group of 3 modes occurs. Closely spaced lie $E_1(LO)$, $A_1(LO)$, and $B_1(high)$. Similarly, the triple group $E_2(high)$, $E_1(TO)$, and $A_1(TO)$ appears between 370 cm^{-1} and 440 cm^{-1}. Finally, the single eigenfrequencies 100 cm^{-1} and 250 cm^{-1} belong to the $E_2(low)$ and

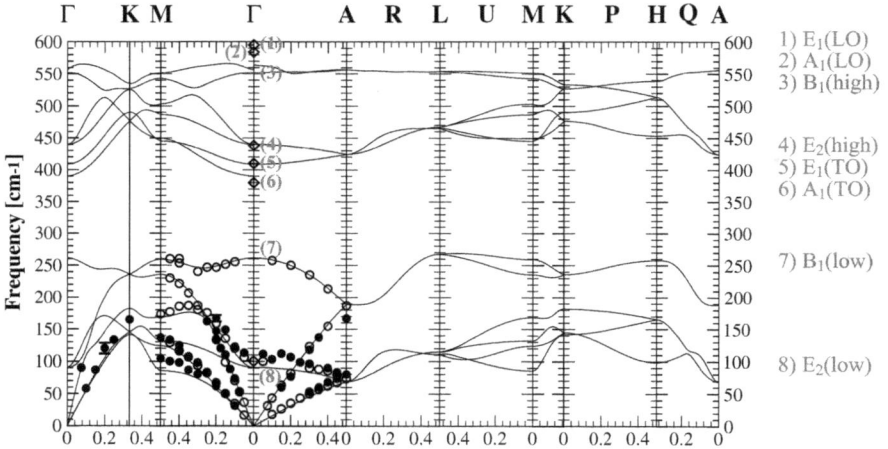

Figure 3.2: *Phonon dispersion of wurtzite ZnO from [Serrano 2003]. The experimental results were derived from Raman scattering (diamond symbols in the BZ center, [Serrano 2003]) and from inelastic neutron scattering (circles, [Hewat 1970, Thoma 1974]). They are well described by calculated results, represented by the solid lines, which were obtained by ab initio calculations. The zone center optical phonon modes A_1, E_1, and E_2 (red) can be observed by Raman scattering, while the B_1 modes (green) are silent.*

the B_1(low) mode, respectively.

For the understanding of the ZnO eigenmode assignment, it is instructive to consider how the phonon eigenmodes of the wurtzite ZnO lattice are related to those of cubic zinc-blende crystals, e.g. ZnSe. In the zinc-blende structure, the cubic symmetry implies the equivalence of the three perpendicular spatial directions. This results in a threefold-degenerate optical phonon mode (F-symmetry). This triple optical phonon is frequency-split into one LO and two TO modes, the frequency of the former exceeding the TO values due to the macroscopic electric field, which results from the bond polarity. Thus, the zinc-blende phonon dispersion curve has two optical phonon frequencies in the Brillouin center, whose difference in frequency is a measure of the bond polarity.

In contrast, the c-axis in the hexagonal wurtzite structure is different from the pair of a- and b-axes, which are symmetrically equivalent. For the optical phonon modes, this implies a symmetry splitting of the above-mentioned triple-degenerate F mode into one mode with an atomic displacement along c (A-symmetry) and a degenerate pair of modes, whose atomic displacement is within the a,b-plane (E-symmetry). The latter are the E_1 modes listed in the representation

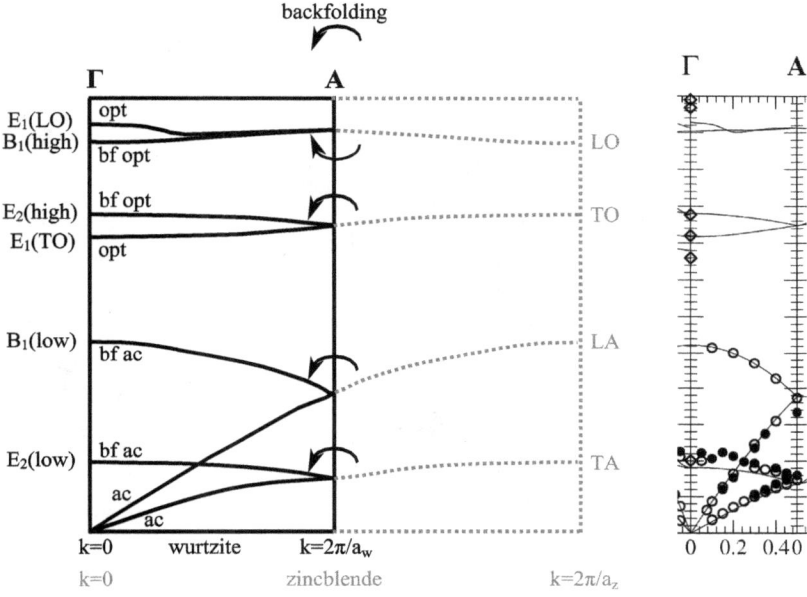

Figure 3.3: *Schematic illustration of the backfolding character of phonon modes in the phonon dispersion relation of wurtzite with respect to the corresponding zinc-blende phonon modes. The doubling of the atomic basis in the real space from zinc blende to wurtzite (4 instead of 2 atoms in the unit cell) corresponds to a bisection of the Brillouin zone in the reciprocal space. For comparison, an excerpt from the ZnO phonon dispersion relation in [Serrano 2003] is shown.*

of Γ_{opt}. The frequency splitting reflects the bond strength anisotropy and is independent of the bond polarity. The polarity induces an additional splitting of the A_1 into $A_1(LO)$ and $A_1(TO)$ as well as of the E_1 modes into $E_1(LO)$ and $E_1(TO)$. As shown in Figure 3.2, the LO-TO frequency splitting for ZnO of about 200 cm^{-1} exceeds by far the value of the mode splitting between A_1 and E_1 (< 30 cm^{-1}). This is due to the pronounced bond polarity (subsection 3.1.1), which distinctly exceeds the bond strength anisotropy. A further consequence of this strong LO-TO splitting and the rather small frequency difference between the A_1 and the E_1 modes is that mode mixing can only occur between $A_1(TO)$ and $E_1(TO)$ (quasi-TO mode) or between $A_1(LO)$ and $E_1(LO)$ (quasi-LO mode) [Bergman 1999]. Such mode mixing is possible in uniaxial crystals for polar phonons with propagation neither parallel nor perpendicular to the c-axis [Loudon 1964]. The frequencies of the mixed modes lie between the corresponding original modes.

The six additional modes of Γ_{opt}, i.e. $2E_2$ (twofold degenerate) and $2B_1$, have no counterpart in the zinc-blende Brillouin zone center. They may be regarded as a result of the backfolding of

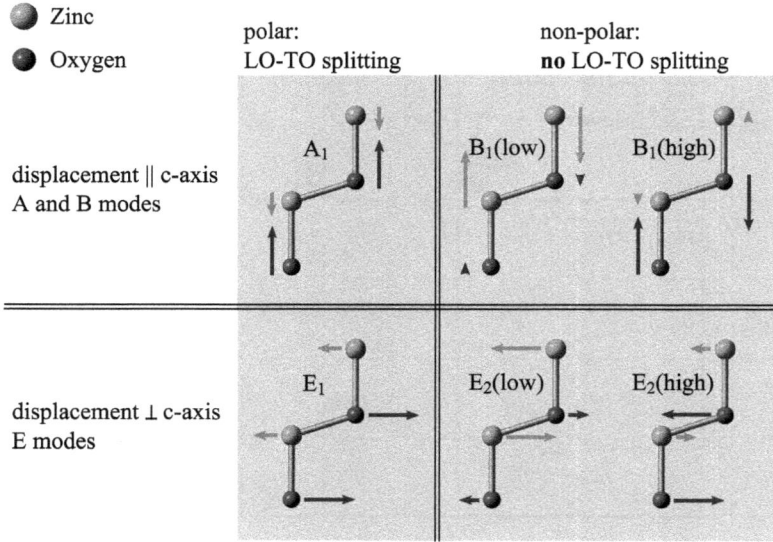

Figure 3.4: *Optical phonon modes of wurtzite ZnO. The atomic displacements are labeled for the four atoms of the unit cell shown in Figure 3.1. The length of the arrows corresponds to the phonon eigenvector values derived for the respective atom sort by DFT calculations in [Serrano 2004]. The polar phonon modes A_1 and E_1 split into LO and TO. The E modes with displacements perpendicular to the c-axis are twofold degenerate.*

the zinc-blende zone-edge modes (LA, TA (2x), LO, and TO (2x)) into the zone center, when turning from zinc blende to wurtzite. This backfolding directly results from the doubling of the atomic periodicity along the wurtzite c-axis with respect to the direction in zinc blende (4 basis atoms instead of 2), corresponding to a bisection in the reciprocal space. The backfolded optical modes are referred to as B_1(high) and E_2(high), whereas the backfolded acoustic modes are called B_1(low) and E_2(low). The backfolding of these modes appears strikingly clear in Figure 3.3 when following the corresponding dispersion curves from the zone center Γ to the zone edge A, i.e. along the c-direction of the wurtzite lattice, and subsequently back to the Γ-point. All these modes are non-polar and therefore exhibit no LO-TO frequency splitting.

In summary, the occurrence of $\Gamma_{opt} = A_1 + 2B_1 + E_1 + 2E_2$ with 9 optical modes in the wurtzite lattice compared to 3 in zinc blende may be explained in terms of (i) an anisotropy-induced splitting of the F-mode from the zinc-blende Γ-point, leading to A_1 and E_1, and (ii) a backfolding of the zinc-blende zone-edge modes, resulting in $2B_1$ and $2E_2$.

Phonons in the wurtzite symmetry are fully characterized by the motion of the four basis atoms [Tsuboi 1964]. The corresponding atomic displacements within the unit cell are shown in Figure 3.4. For the A and B modes, the displacements are directed along the c-axis, and they are distinct in the following way: The A-mode pattern consists of an oscillation of the rigid sublattices, Zn versus O. Due to the bond polarity, this results in an oscillating polarization. For the B modes, in contrast, one sublattice is essentially at rest, while in the other sublattice the neighboring atoms move in opposite directions. In the case of the B_1(low) mode, the heavier Zn sublattice is distorted, while the B_1(high) involves the lighter O sublattice. No net polarization is induced by the B modes. Thus, the A and B modes may be classified as one polar and two non-polar modes. The same scheme applies for the E modes with their atom displacement perpendicular to the c-axis. As stated before, the E modes are twofold degenerate, because the two axes perpendicular to the c-axis are energetically equivalent, even though linearly independent. The E_1 mode is an oscillation of rigid sublattices and consequently exhibits an oscillating polarization. In contrast, the E_2 modes, E_2(low) and E_2(high), with essentially one rigid sublattice and the other one oscillating in itself, are non-polar. Thereby, the low-wavenumber E_2(low) mode predominantly involves the vibration of the heavy Zn sublattice, while the high-wavenumber E_2(high) mode is mainly associated with the vibration of the lighter O sublattice.

As discussed in subsection 2.1.1, generally only optical modes at the center of the Brillouin zone are candidates for Raman scattering. Additionally, the Raman selection rules have to be fulfilled, i.e. phonon modes must have non-vanishing components in the Raman tensor to be Raman active (subsection 2.1.3):

$$\Re = \begin{pmatrix} \Re_{xx} & \Re_{xy} & \Re_{xz} \\ \Re_{yx} & \Re_{yy} & \Re_{yz} \\ \Re_{zx} & \Re_{zy} & \Re_{zz} \end{pmatrix} . \tag{3.1}$$

While the B_1 modes are silent, the modes with A_1-, E_1-, and E_2-symmetry are Raman active with the following Raman tensors [Cusco 2007, Arguello 1969]:

$$\Re(E_2^{(1)}) = \begin{pmatrix} d & 0 & 0 \\ 0 & -d & 0 \\ 0 & 0 & 0 \end{pmatrix} ; \qquad \Re(E_2^{(2)}) = \begin{pmatrix} 0 & d & 0 \\ d & 0 & 0 \\ 0 & 0 & 0 \end{pmatrix} ;$$

$$\Re(E_1(x)) = \begin{pmatrix} 0 & 0 & c \\ 0 & 0 & 0 \\ c & 0 & 0 \end{pmatrix} ; \qquad \Re(E_1(y)) = \begin{pmatrix} 0 & 0 & 0 \\ 0 & 0 & c \\ 0 & c & 0 \end{pmatrix} ;$$

$$\Re(A_1(z)) = \begin{pmatrix} a & 0 & 0 \\ 0 & a & 0 \\ 0 & 0 & b \end{pmatrix} , \qquad\qquad (3.2)$$

where a, b, c, d are material constants and the coordinates x, y, z in parentheses describe the phonon polarization of the polar modes in a laboratory coordinate system (x,y,z) with z-axis parallel to the c-axis of the wurtzite ZnO. Whether these modes are observable in a particular experiment or not, depends on the scattering configuration as described in the Porto notation $\vec{k}_i(\vec{\Pi}_i, \vec{\Pi}_s)\vec{k}_s$ (subsection 2.1.3). In this notation, 180° backscattering with the incident light along the c-axis and the incident and detected polarization parallel is given by $z(xx)\bar{z}$. Figure 3.5 displays which wurtzite ZnO phonons are observed by Raman scattering in different configurations. In the adjoining table, the optical phonon modes and their configuration requirements as well as Raman shifts are summarized. Note that in the Raman spectra E_2 modes are also observed in forbidden configurations due to non-perfect crystal alignment or quality. Because of the strong occurrence of E_2 modes in standard backscattering experiments, they can be considered as a Raman fingerprint for ZnO. Besides the principal optical phonons, a rather strong multi-phonon mode at about 330 cm^{-1} can be seen in various configurations. By temperature-dependent measurements it can be identified as a difference mode (subsection 4.1). Recently, it could be assigned to the process E_2(high)-E_2(low) by symmetry considerations [Cusco 2007].

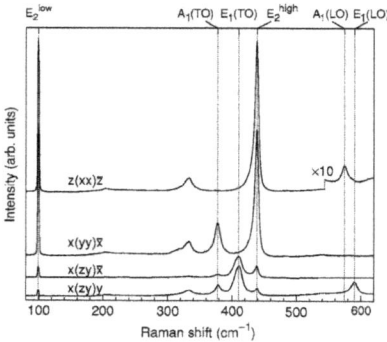

Config.	Modes	Raman shift (cm^{-1})
$z(xx)\bar{z}$	A_1(LO), E_2(h), E_2(l)	574, 438, 99
$x(yy)\bar{x}$	A_1(TO), E_2(h), E_2(l)	378, 438, 99
$x(zy)\bar{x}$	E_1(LO)	410
$x(zy)y$	E_1(LO), E_1(TO)	590, 410

Figure 3.5: *Allowed optical phonon modes in the Raman spectra of wurtzite ZnO for different experimental configurations [Cusco 2007].*

The micro-Raman experiments conducted for this thesis imply backscattering configuration and also all macro-Raman experiments were conducted in backscattering geometry. As the surface of most of the studied ZnO systems with well defined orientation is c-plane-oriented (0001), the scattering configurations valid for most experiments in this thesis are $z(xx)\bar{z}$ and $z(xy)\bar{z}$. Possible relaxation of the Raman selection rules can be caused by disordered or not perfectly aligned crystals (subsection 2.3.2). Additionally, as stated in subsection 2.2.2, in micro-Raman scattering the light is not solely collected from the 180° backscattering angle but from a finite solid angle. Furthermore, samples like wet-chemically synthesized ZnO nanoparticles or polycrystalline VPT ZnO systems do not possess a macroscopically defined orientation. Thus, the corresponding Raman scattering results reflect all possible orientations.

By very sensitive Raman experiments, Cusco et al. observed a variety of additional Raman signals and assigned them to multi-phonon modes by their symmetry and temperature behavior [Cusco 2007]. For a complete overview see Table 3.1. As the individual participating phonons are not restricted to q = 0 in multi-phonon scattering, various features can be expected in the higher wavenumber region. They reflect the multi-phonon density of states, but still the multi-phonon selection rules have to be fulfilled [Siegle 1997]. Figure 3.6 displays phonon density of state curves calculated using ab initio methods [Serrano 2004]. The one-phonon density of states ranges from 0 cm^{-1} to about 600 cm^{-1}, with a gap between the acoustic and optical branches from 270 cm^{-1} to 370 cm^{-1}. This gap is important for the possibility of impurity gap modes (see subsection 2.3.2). Local modes are possible for frequencies above the optical mode branches, i.e. above 600 cm^{-1}. Note that the corresponding energy is rather high and only impurities lighter than oxygen can be expected to induce local modes. While the two-phonon difference modes are obviously located in the frequency range of the one-phonon branches, the frequency range above 570 cm^{-1} is dominated by two-phonon sum modes and multi-phonon processes of higher order. The intensity of multi-phonon processes is particularly strong in resonant Raman scattering [Calleja 1977] when the exciting light approaches the band gap energy.

3.1.3   Band gap and optical properties

ZnO possesses a large band gap of 3.4 eV. The conduction band minimum, formed by empty 4s orbitals of Zn^{2+} (or the antibonding sp^3 hybrid states), and the valence band maximum, formed by occupied 2p orbitals of O^{2-}(or the bonding sp^3 hybrid states), both lie at the center of the Brillouin zone (Γ-point) [Klingshirn 2007], i.e. the band gap of ZnO is direct. In Figure 3.7, the band structure at the Γ-point is shown. Due to crystal-field and spin-orbit interaction, the valence band is split into three states (A, B, C). The upper valence subband (A) and the conduction band both have Γ_7 character [Meyer 2004]. The large band gap energy of 3.4 eV corresponds to a wavelength of 365 nm, i.e. in the UV. Therefore, high-quality ZnO is highly

Process	Raman shift (cm^{-1})	Brillouin zone points / lines
E_2(low)	99	Γ
2TA; $2E_2$(low)	203	L, M, H ;Γ
B_1(high)-B_1(low)	284	Γ
E_2(high)-E_2(low)	333	Γ
A_1(TO)	378	Γ
E_1(TO)	410	Γ
E_2(high)	438	Γ
2LA	483	M-K
$2B_1$low; 2LA	536	Γ; L, M, H
A_1(LO)	574	Γ
E_1(LO)	590	Γ
TA+TO	618	H, M
TA+LO	657	L, H
TA+LO	666	M
LA+TO	700	M
LA+TO	723	L-M
LA+TO	745	L-M
LA+TO	773	M, K
LA+TO	812	L, M
2TO	980	L-M-K-H
TO+LO	1044	A, H
TO+LO	1072	M, L
2LO	1105	H, K
$2A_1$(LO),$2E_1$(LO); 2LO	1158	Γ; A-L-M

Table 3.1: *Raman active phonons and phonon combinations of wurtzite ZnO and their wavenumber values from [Cusco 2007]. In the third column, the corresponding points or lines of the processes in the Brillouin zone are denoted.*

transparent throughout the visible spectral range. Moreover, the wavelength of the exciting laser in resonance Raman experiments is required to be in the UV and the experimental equipment must be UV-compatible. On the other hand, Raman spectroscopy with excitation in the visible spectral range has the advantage of a large scattering volume due to the transparency of ZnO. By alloying ZnO with MgO (band gap 7.5 eV) or CdO (band gap 2.3 eV), the fundamental band gap can be tailored to the particular application within a large energetic range.

The optical properties of ZnO are strongly influenced by the electronic band structure and the phononic properties. In Figure 3.7 a photoluminescence (PL) spectrum of n-type bulk ZnO with several characteristics is shown. The (near-band-gap) excitonic emission and donor-acceptor pair (DAP) emission with their optical phonon replica dominate the PL in the high energy region. Additionally, the so-called green band between about 440 nm (2.8 eV) and 650 nm (1.9 eV) occurs, which is generally attributed to impurities or defects in ZnO [Klingshirn 2007].

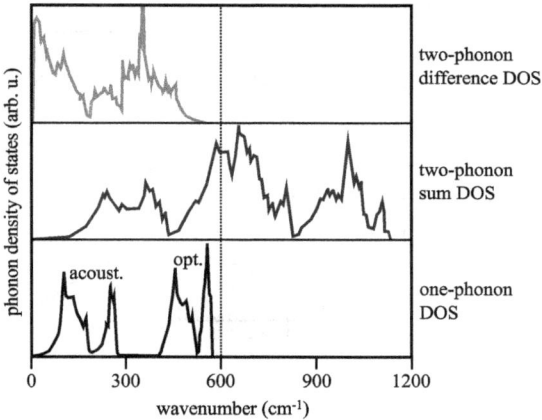

Figure 3.6: *One-phonon, two-phonon-sum, and two-phonon-difference density of states, redrawn from [Serrano 2004]. Note the frequency gap between the acoustic and optical phonon branches from 270 cm^{-1} to 370 cm^{-1} in the one-phonon density of states.*

3.2 Growth, processing, and applications

As mentioned before, ZnO is already used in large-scale industrial manufacturing (food additives, cosmetics, plastics, etc.). Due to progress in processing and theoretical understanding of its unique optical, semiconducting, and piezoelectric properties, conventional applications can be improved and, additionally, new application potentials arise. For example, ZnO-based systems could represent an important light source in the future or be utilized as a transparent conducting oxide for applications such as front electrodes in solar cells and liquid crystal displays. The major technical challenge for many of the new applications is to achieve stable and reproducible p-type ZnO. Theoretical aspects and the experimental situation of the p-doping issue are reviewed in subsection 3.2.1. Raman scattering results on nitrogen-doped ZnO systems are presented and discussed in chapter 6.

ZnO has also been considered for spintronics applications since theoretical predictions of room temperature ferromagnetism (RT FM) in ZnMnO. While experimental evidence for such RT FM has been reported for ZnO alloyed with different transition metals (TM), it stays unclear whether the observed magnetic properties are intrinsic, i.e. due to substitutional incorporation of the magnetic impurities, or if they result from either magnetic secondary phases or experimental negligences. The theory of the diluted magnetic semiconductor ZnO and experimental findings are summarized in subsection 3.2.2. In chapter 5, experimental results on ZnO:TM systems are presented for a variety of transition metals. As new applications of ZnO often involve the

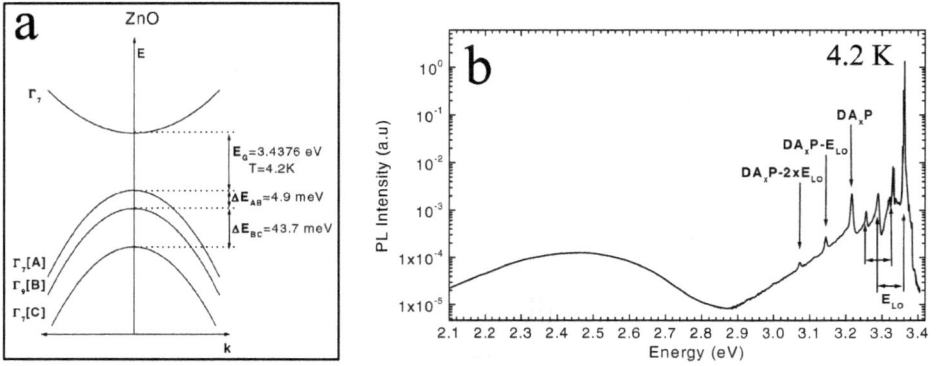

Figure 3.7: *(a) ZnO band structure at the Γ-point of the Brillouin zone [Meyer 2004]. The conduction band (empty Zn^{2+} 4s orbitals) and the highest valence subband (occupied O^{2-} 2p orbitals) possess Γ_7 symmetry. (b) Photoluminescence of bulk n-type ZnO [Meyer 2004]. The spectrum is dominated by the so-called green band (from impurities or defects), and, in the blue to UV spectral range, by excitonic and donor-acceptor pair emission with the corresponding phonon replica.*

incorporation of impurities, Figure 3.8 gives an overview over the elements which make potential candidates or are already used for p- or n-doping, band gap engineering, or magnetic alloying. There are manifold growth techniques for ZnO. Most of the samples for this thesis were fabricated using molecular beam epitaxy, vapor phase transport, wet-chemical synthesis, or hydro-thermal growth. Impurity incorporation was carried out during growth or via post-growth ion implantation. Detailed growth parameters are given in the results sections, and ion implantation is discussed in subsection 4.1.1 and section 5.1. Another major trend in ZnO research is the fabrication of ZnO nanosystems with various morphologies and great potential for future application, see subsection 3.2.3. Raman spectroscopic results on ZnO nanostructures are presented in section 4.2.

3.2.1   Doping of ZnO

The doping of ZnO and the influence of various impurities and intrinsic defects in ZnO are reviewed in detail in [Look 2006]. As-grown ZnO is generally n-type with dominating shallow donors. This n-type character was found to be due to hydrogen impurities and intrinsic defects. While the oxygen vacancy V_O and the zinc interstitial Zn_I indeed act as donors, they are not expected to play a major role due to their high formation energy [Kohan 2000]. Other native-

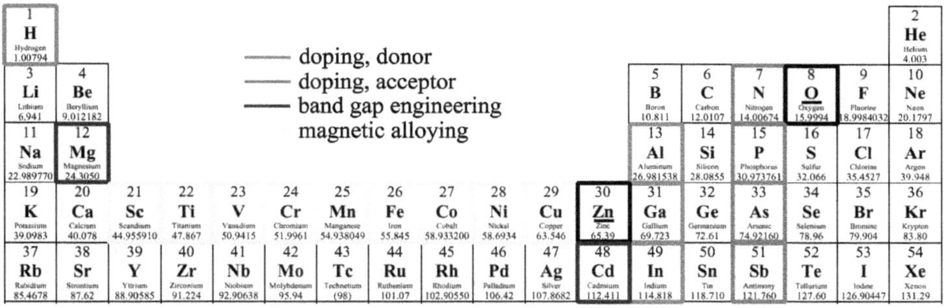

1 H Hydrogen 1.00794																	2 He Helium 4.003
3 Li Lithium 6.941	4 Be Beryllium 9.012182											5 B Boron 10.811	6 C Carbon 12.0107	7 N Nitrogen 14.00674	8 O Oxygen 15.9994	9 F Fluorine 18.9984032	10 Ne Neon 20.1797
11 Na Sodium 22.989770	12 Mg Magnesium 24.3050											13 Al Aluminum 26.981538	14 Si Silicon 28.0855	15 P Phosphorus 30.973761	16 S Sulfur 32.066	17 Cl Chlorine 35.4527	18 Ar Argon 39.948
19 K Potassium 39.0983	20 Ca Calcium 40.078	21 Sc Scandium 44.955910	22 Ti Titanium 47.867	23 V Vanadium 50.9415	24 Cr Chromium 51.9961	25 Mn Manganese 54.938049	26 Fe Iron 55.845	27 Co Cobalt 58.933200	28 Ni Nickel 58.6934	29 Cu Copper 63.546	30 Zn Zinc 65.39	31 Ga Gallium 69.723	32 Ge Germanium 72.61	33 As Arsenic 74.92160	34 Se Selenium 78.96	35 Br Bromine 79.904	36 Kr Krypton 83.80
37 Rb Rubidium 85.4678	38 Sr Strontium 87.62	39 Y Yttrium 88.90585	40 Zr Zirconium 91.224	41 Nb Niobium 92.90638	42 Mo Molybdenum 95.94	43 Tc Technetium (98)	44 Ru Ruthenium 101.07	45 Rh Rhodium 102.90550	46 Pd Palladium 106.42	47 Ag Silver 107.8682	48 Cd Cadmium 112.411	49 In Indium 114.818	50 Sn Tin 118.710	51 Sb Antimony 121.760	52 Te Tellurium 127.60	53 I Iodine 126.90447	54 Xe Xenon 131.29

Legend:
—— doping, donor
—— doping, acceptor
—— band gap engineering
—— magnetic alloying

Figure 3.8: *Periodic table of elements. Labeled are elements which make potential candidates or are already used for p- or n-doping, band gap engineering, or magnetic alloying of ZnO.*

defect donors, especially complexes involving Zn_I, may have a more important influence on the n-type character of ZnO [Look 2005]. The dominant donor in as-grown ZnO is hydrogen, which is always a donor in ZnO and has a low formation energy [Van de Walle 2000]. Besides the n-type properties of the as-grown ZnO, also additional n-doping proves to be uncomplicated, as the group III elements Al, Ga, and In can easily be incorporated on Zn sites, resulting in possible carrier concentrations beyond 10^{20} cm^{-3}.

While n-type ZnO is obviously available without complications, p-doping is probably the primary technical issue of ZnO processing. Low acceptor concentrations in as-grown ZnO can be explained by zinc vacancies V_{Zn}, but are clearly overcompensated by the presented n-type properties. While the group I elements Li and Na are expected to be shallow acceptors on Zn sites, both can be incorporated as donors (Li_I, Na_I) as well. Thus, they exhibit a too strong self-compensation. Among the candidates for impurities acting as acceptors, the most promising are the group V elements N, P, As, and Sb. Nitrogen on oxygen sites (N_O) is a shallow acceptor in ZnO with a binding energy of about 100 meV. Regarding its ionic size, (N_O) should be the ideal acceptor. While indeed high N_O concentrations of up to 10^{20} cm^{-3} were demonstrated, in most cases the active acceptor concentration is much lower, most probably due to passivation by hydrogen. Possible remedies against the acceptor passivation by hydrogen are post-growth thermal annealing or sample irradiation. Consequentially, successful p-type ZnO via nitrogen incorporation and subsequent fabrication of blue ZnO-based LEDs was achieved using a laborious, temperature-modulated MBE growth [Tsukazaki 2004]. Other promising experiments towards p-doping of ZnO have been conducted with the other group V elements P, As, and Sb. Still, no straightforward procedure has been established for fabricating reproducible and stable p-type ZnO of high quality.

3.2.2   ZnO:TM as DMS system

In the field of spintronics (spin transport electronics), researchers investigate the spin degree of freedom of electrons with respect to its application either in conventional electronics or in a new, solely spin-based technology with expected advantages such as non-volatility, increased data processing speed, decreased electric power consumption, and increased integration densities [Wolf 2001]. Though, much in this area is speculative and there are also skeptic opinions as to the predicted superiority of spintronics compared to conventional electronics [Bandyopadhyay 2004]. Major technical issues are yet to be solved in order to realize a fully functional spin-based technology, e.g. efficient spin injection and controlled spin transport. Furthermore, potential materials for spintronics have to meet various requirements including a high spin-polarization. A key material system for such applications could be the class of diluted magnetic semiconductors (DMS). In DMS, non-magnetic host ions are partially substituted by magnetic ions, most frequently by transition metal (TM) ions [Furdyna 1988]. While a major advantage of such systems would be their great potential of combining spintronics with conventional semiconductor-based electronics, it is unclear whether DMS can be fabricated which meet the material requirements. Well-understood DMS systems already exist, e.g. GaMnAs, with very promising magnetic properties, however, with Curie temperatures (T_C) below room temperature, which strongly limits their application potential.

Using a mean-field Zener model, Dietl et al. presented two promising DMS candidates with predicted stable ferromagnetic configurations above room temperature arising from carrier-mediated exchange interaction [Dietl 2000]: GaMnN and ZnMnO. However, according to these calculations, p-type ZnO is required as host material. While Mn acts as an acceptor in the III-V compound GaMnAs, it is isoelectric in ZnO. As discussed in subsection 3.2.1, high quality p-type ZnO is not easily available (yet). This is even more problematic when ZnO is alloyed with TM ions because the n-type character of ZnO due to intrinsic defects is increased by the disordering. Ab initio calculations again showed stable room temperature ferromagnetism (RT FM) for p-type ZnO alloyed with Mn, but additionally for n-type ZnO alloyed with other TM ions [Sato 2001]. Furthermore, a model for ferromagnetic coupling in n-type diluted oxides due to bound magnetic polarons was proposed in [Coey 2005].

In the following paragraphs, the mentioned theoretical studies will be outlined. Finally, the experimental situation with its ambiguous and often contradictory results regarding the magnetic properties of different $Zn_{1-x}TM_xO$ systems is reviewed.

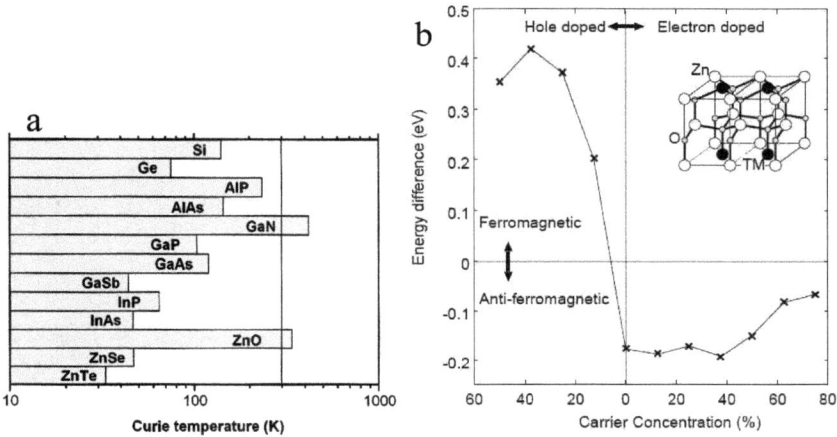

Figure 3.9: *(a) According to a mean-field Zener model, ZnO and GaN with 5% Mn and a hole concentration of 3.5 x10^{20} cm^{-3} are promising materials for ferromagnetism at room temperature [Dietl 2000]. (b) According to ab initio calculations, p-type host ZnO is required for a stable ferromagnetic configuration of ZnMnO [Sato 2001].*

Mean-field Zener model from [Dietl 2000]

Dietl et al. presented a mean-field Zener model for Mn-alloyed semiconductors, which successfully describes magnetic properties of p-type GaMnAs and ZnMnTe [Dietl 2000]. In this model, the direct interaction between the half-filled 3d shells of adjacent Mn atoms leads to an antiferromagnetic configuration (super exchange). The ferromagnetic correlation, on the other hand, arises from interactions of the localized Mn spins mediated by free holes from shallow acceptors in doped magnetic semiconductors. Mn in GaMnAs is, as discussed before, a localized spin as well as an acceptor, whereas II-VI semiconductors have to be doped additionally to fulfill the requirements of this model.

In this theoretical framework, the Ginzburg-Landau free-energy functional is expressed as a function of the magnetization by the localized spins and then minimized. The corresponding Curie temperature is calculated using a mean-field approximation for the long-range exchange interactions. The T_C value results from a competition between the antiferromagnetic direct Mn-Mn interaction and the hole-mediated ferromagnetic coupling and it depends on material parameters, Mn concentration, and hole density. Encouraged by successful results for GaMnAs and ZnMnTe, additional values were computed for various semiconductors containing 5% Mn and 3.5 x10^{20} cm^{-3} holes. The highest Curie temperatures ($T_C > RT$) were calculated for GaMnN and

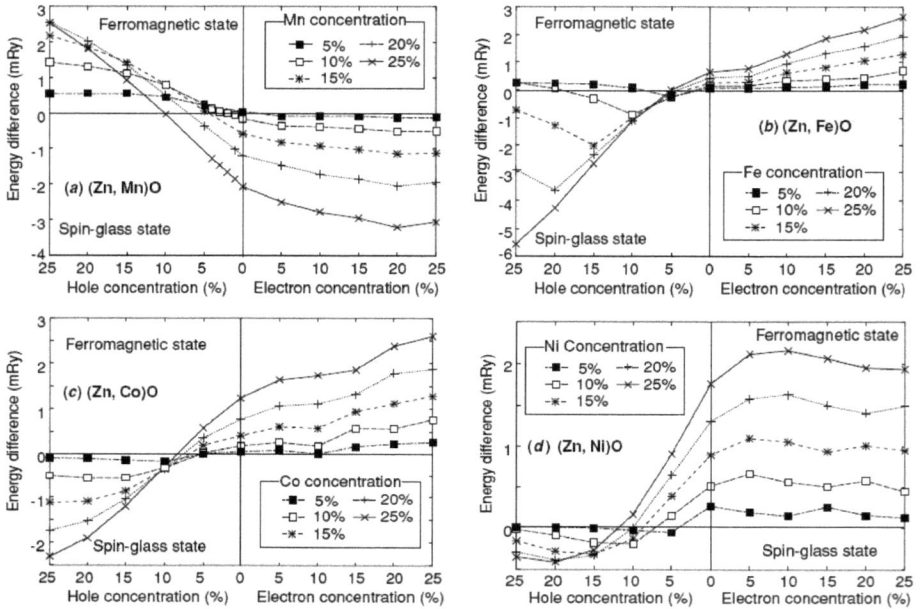

Figure 3.10: *The plots from [Sato 2002] show the calculated energy difference between the ferromagnetic state and the spin-glass state versus the carrier concentration in ZnTMO. The results are shown for different transition metals and transition metal concentrations. While p-type host ZnO is required for a ferromagnetic configuration of ZnMnO (a), n-type host ZnO is more promising for Fe, Co, and Ni (b-d).*

ZnMnO, see Figure 3.9a. While ZnMnO is thereby identified as a promising DMS candidate, the limitations of the model have to be kept in mind. The results are only valid for the transition metal Mn and for p-type ZnO with a high carrier concentration.

Ab initio calculations from [Sato 2001, Sato 2002]

Sato et al. conducted ab initio calculations on the electronic structure of TM-alloyed ZnO [Sato 2001], using a Green's-function method based on the local spin density approximation (LSDA). In a wurtzite supercell with 8 ZnO molecules, two Zn atoms were substituted by TM ions, corresponding to a TM concentration of 25%. Since TM substitution of Zn is isoelectric, doping was induced by additional substitutions: N_O for p-doping and Ga_{Zn} for n-doping (see subsection 3.2.1). By this approach, the electronic structure is calculated for the ferromagnetic state (parallel spins) and the antiferromagnetic state (partly antiparallel spins). The energy difference $\Delta E = E_{afm}\text{-}E_{fm}$ then gives the more stable configuration. The results are presented in Figure 3.9b: In the case of p-type ZnMnO, ferromagnetic ordering is more stable, while for n-type or insulating ZnMnO, antiferromagnetism can be expected.

While this result is in accordance with the findings in [Dietl 2000], an analysis of the total DOS in [Sato 2001] suggests a different mechanism for the ferromagnetic coupling in p-type ZnMnO: Carrier-hopping between partially occupied 3d-orbitals of the TM impurities leads to a ferromagnetic alignment of neighboring ions, i.e. the so-called double exchange. In the case of Mn with its half-filled 3d shells, electron doping does not yield the expected stabilization of such a ferromagnetic configuration. In contrast, according to the ab initio calculations, such stabilization by donors can be expected in ZnO alloyed with Ti, V, Cr, Fe, Co, Ni, and Cu. In [Sato 2002], the stability of the ferromagnetic ordering was calculated by the same method for a variety of carrier and TM concentrations. For Ni, Fe, and Co indeed n-type ZnO was found to be the more promising host material (see Figure 3.10).

Findings of [Sato 2001] regarding undoped ZnO as host material were challenged by the work of [Spaldin 2004], where, according to similar LSDA-based DFT calculations, p-type ZnO was found to be necessary for stable ferromagnetism in both Mn- and Co-alloyed ZnO. It should also be noted that the calculations in [Sato 2001, Sato 2002] require very high TM concentrations of up to 25% for $T_C > RT$, depending on the TM.

Donor impurity band exchange model from [Coey 2005]

While all above-mentioned theoretical findings agree in the fact that n-type Mn-alloyed ZnO should show no ferromagnetism, several experimental results report RT FM for this system (see next paragraph and Table 3.2). A model was proposed in [Coey 2005] in order to explain ferromagnetic exchange coupling in n-type diluted magnetic oxides, including ZnO. In this model, a material system is considered where magnetic atoms substitute non-magnetic cations and, additionally, donor defects are present (e.g. oxygen vacancies in ZnO, see subsection 3.2.1). For a sufficient donor defect concentration, the hydrogenic orbitals of the electrons associated with the defects can overlap to form a delocalized impurity band. The donors then tend to form bound magnetic polarons, coupling the 3d moments of the magnetic ions within their orbits. This mechanism is illustrated in Figure 3.11. Generally, the coupling between a donor electron and a magnetic cation is ferromagnetic if the 3d shell is less than half full, but antiferromagnetic otherwise. The mediated coupling between the magnetic ions themselves is ferromagnetic either way. Long-range ferromagnetic order, finally, occurs above the percolation thresholds of both the polaron and the magnetic cation concentration.

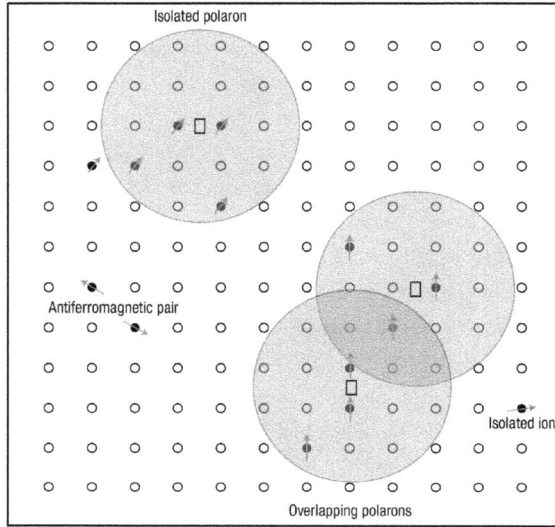

Figure 3.11: *Figure from [Coey 2005]: Donor electrons in ZnO (here: due to oxygen vacancies) tend to form bound magnetic polarons, which couple the 3d moments of the magnetic ions (arrows) within their orbits. Zinc atoms are represented by circles and oxygen vacancies by squares, while the regular oxygen sublattice is not shown.*

Note that this model is applicable both for n-type and p-type material and that the mechanism already works for a comparatively low carrier concentration.

Experimental situation

Since the theoretical works in [Dietl 2000] and [Sato 2001], an extensive research activity was resulting in numerous publications on transition-metal-alloyed ZnO in the past years. While in several of these studies an intrinsic ferromagnetism of ZnTMO is stated [Sharma 2003, Ueda 2001], the entire experimental situation advises caution. Systems with many different parameters have been studied, which makes a clear picture difficult. Among the studied systems are ZnO nanostructures, bulk crystals, thin films, and ceramics, alloyed with different transition metals and transition metal combinations, and fabricated using various different growth techniques and after-growth treatments. Furthermore, results on the magnetic properties of TM-alloyed ZnO are often contradictory, even for similar samples and experimental conditions. In Table 3.2, prominent experimental studies regarding the magnetic properties of transition-metal-alloyed

Sample	TM	RT FM	Origin	Author
film	V	yes	intrinsic	[Saeki 2001]
film	Co	yes	intrinsic	[Ueda 2001]
film	Cr,Mn,Ni	no	-	[Ueda 2001]
film/powder	Co	yes/no	intrinsic/-	[Lee 2002]
nanorods	Co	yes	sec.phase	[Ip 2003]
film	Co	yes	sec.phase	[Norton 2003]
bulk/film	Mn	yes	intrinsic	[Sharma 2003]
film	Co	yes	sec.phase	[Park 2004]
film	Sc,Ti,V,Fe,Co,Ni	yes	intrinsic	[Venkatesan 2004]
film	Cr,Mn,Cu	no	-	[Venkatesan 2004]
film	Cu	yes	intrinsic	[Buchholz 2005]
nanocrys.	Mn,Co	yes	intrinsic	[Kittilstved 2005]
bulk	Mn,Co	no	-	[Lawes 2005]
film	Mn	yes	substr.	[Mofor 2005]
powder	Mn,Co	no	-	[Rao 2005]
polycrys.	Fe+Cu	yes	sec.phase	[Shim 2005]
nanowires	Mn	yes	intrinsic	[Philipose 2006]
film	Mn,Co	yes	ZnO def.	[Hong 2007]
nanocrys.	Fe	yes	intrinsic	[Karmakar 2007]
film	Cu	yes	sec.phase	[Sudakar 2007]
film	Mn	yes	intrinsic	[Xu 2007]

Table 3.2: *Studies on the magnetic properties of transition-metal-alloyed ZnO. The third and the fourth column denote whether room temperature ferromagnetism was observed and what origin was identified for this RT FM by the authors, respectively.*

ZnO are presented. In accordance with other reviews of the experimental situation [Liu 2005, Norton 2006, Ozgur 2005, Seshadri 2005], the results are ambiguous with respect to the existence and origin of a RT FM in ZnO-based DMS systems.

In addition to the difficulties of providing high-quality p-type ZnO, often extrinsic effects cannot be sufficiently excluded as cause for the observed properties. In [Mofor 2005], for example, ferromagnetism in ZnMnO layers was identified as mainly extrinsic, originating from slightly magnetic substrate material. Furthermore, magnetic contamination of samples due to handling with stainless-steel tweezers was found in [Abraham 2005]. Still, the key issue is the formation of secondary phases because the required transition metal concentrations are typically in the range of 5-25% and therefore often near or above the corresponding solubility limits in ZnO [Jin 2001, Kolesnik 2004]. All theoretical models, however, demand substitutional incorporation of the TM ions on Zn sites. Since not only the elemental TM clusters, but also most TM oxides show distinct magnetic properties, minor precipitate formation may already dominate the magnetic properties of such samples.

To summarize, up to now the magnetic properties of ZnTMO systems are not fully understood, and the often contradicting experimental results require further thorough research. Especially, microstructural studies should always be included to evaluate the influence of magnetic secondary phases. In chapter 5, Raman spectroscopy is successfully applied to study the impact of TM impurity incorporation on the ZnO crystal and to identify magnetic secondary phase inclusions.

3.2.3 ZnO-based nanostructures

Much of the future potential of ZnO lies in nanostructured ZnO, for instance in nanolasers, -sensors, -resonators, -cantilevers, and -field-effect-transistors [Wang 2004] as well as in many other mechanical, electronic, photonic, and biomedical applications. Specific physical properties of ZnO nanostructures due to size effects are reviewed in [Ozgur 2005, Wang 2004]. Because of the strong tendency of ZnO to self-organized growth, nanostructures of various different morphologies can be grown by straightforward fabrication techniques: Nanoparticles (quasi-0D), -wires/-rods (quasi-1D), -belts, -tubes, -cages, and many more. Important growth techniques for such nanostructures include MBE or pulsed laser deposition (PLD). Another especially versatile fabrication method is the solid-vapor-phase technique [Wang 2004]. In this process, the source material is vaporized in a furnace and condenses on a substrate. Thereby, different morphologies are achieved by variation of the growth parameters, such as temperature, carrier gas, substrate, and source material. Examples of various morphologies grown by this method are shown in Figure 3.12.

Nanorods/-wires are successfully grown by the vapor-liquid-solid (VLS) approach [Wang 2004]. In this growth process, metal droplets with diameters in the nanometer range (corresponding to the desired rod/wire diameter) serve as catalyst in the 1D ZnO growth. The gas phase reactant is absorbed by the liquid droplet and, after supersaturation, the ZnO nucleation starts.

Among the mentioned ZnO nanostructures, this thesis focuses on nanoparticles. The majority of the studied nanoparticles were prepared by wet-chemical synthesis, few by spray pyrolysis. In the latter, precursors are sprayed on a heated substrate, where they react with each other forming nanoparticles. The principles of the wet-chemical synthesis developed for the samples of this thesis are described in [Chory 2007]: A low temperature synthesis from ethanolic solutions results in nanocrystalline ZnO powder with various organic molecules as potential stabilizing

Figure 3.12: *ZnO nanostructures with various morphologies, fabricated with a solid-vapor process [Wang 2004].*

ligands. The synthesis parameters and different synthesis variants used are discussed in detail in section 4.2.

Part II

Results and discussion

Chapter 4

Pure ZnO: bulk crystals, disorder effects, and nanoparticles

In this chapter, pure ZnO systems are investigated by Raman spectroscopic means. Important questions addressed are the structural characteristics of samples with different morphologies or fabricated using different growth processes. In section 4.1, bulk ZnO single crystals are characterized, which act as host crystals for the implanted ZnO systems discussed in the chapters 5 and 6. Using these high-quality ZnO single crystals and, for comparison, structurally inferior, polycrystalline bulk ZnO, general Raman scattering properties of non-ideal ZnO are presented. The impact of ion irradiation on the structural quality of the ZnO host crystals is analyzed in subsection 4.1.1. In section 4.2, finally, ZnO nanostructures are studied, with focus on structural properties and size effects of wet-chemically synthesized nanoparticles.

4.1 ZnO single crystals and polycrystalline ZnO

In the following, hydrothermally grown ZnO single crystals from CrysTec, Berlin, are characterized by Raman spectroscopy. These samples were taken as host crystals for the implantation of ZnO with transition metal ions (magnetic alloying) and nitrogen ions (doping), discussed in the chapters 5 and 6, respectively. As will be shown in the course of this thesis, these single crystals contain residual impurities, for example Mn and Fe. However, the impurity density is far below the Raman detection limit and they do not influence the Raman studies presented in this section. Hence, these single crystals can be considered as model systems for well-ordered, pure ZnO.

Figure 4.1a shows Raman spectra of such a ZnO single crystal, which were recorded in different scattering configurations with well-defined crystal orientation as well as well-defined incident and detected polarization. Like in all Raman experiments for this thesis, the spectra were recorded in 180° backscattering. To describe the respective scattering configuration, the Porto notation is used, which was introduced in subsection 2.1.3. The most commonly used configuration $z(xx)\bar{z}$ corresponds to incident and scattered light directions along the ZnO c-axis as well as to incoming and detected polarization parallel to each other and perpendicular to the ZnO c-axis. The characteristic Raman features for this symmetry are the E_2(low) and E_2(high) phonon modes at about 99 cm^{-1} and 437 cm^{-1}, respectively. The A_1(LO) phonon mode at about 577 cm^{-1} is allowed in this symmetry, but occurs with very weak intensity in this well-ordered ZnO. The even smaller feature at about 537 cm^{-1} corresponds to a 2xLA phonon mode process [Cusco 2007]. An additional feature occurs at about 332 cm^{-1}. This mode was tentatively, and falsely, attributed to a two-phonon sum from outside the BZ center [Calleja 1977], which assignment is still commonly used. However, Figure 4.1b clearly shows that this mode disappears for low temperatures, indicating a difference phonon process as origin. From this behavior and the frequency position, the feature is attributed to the E_2(high)-E_2(low) difference mode. As stated above, the strongest modes in this configuration are the E_2(high) at about 437 cm^{-1} and the E_2(low) at about 99 cm^{-1}. The frequency difference between these strong modes agrees well with the observed wavenumber position of the Raman feature at about 332 cm^{-1}. This assignment as a second order process from the difference E_2(high)-E_2(low) was recently confirmed by symmetry considerations and a temperature-dependent intensity analysis [Cusco 2007].

In most experiments for this thesis, no polarization selection of the scattered light was used in order to be sensitive for diagonal as well as off-diagonal Raman tensor components of possible additional Raman peaks. For the commonly used scattering configuration described above, this corresponds to the detection of both $z(xx)\bar{z}$ and $z(xy)\bar{z}$ contributions. However, no additional bulk modes of pure ZnO are expected from the cross polarization in this case because only the E_2 modes are allowed for $z(xy)\bar{z}$, as can be seen in Figure 4.1a. In experiments with directions of the incident and scattered light perpendicular to the ZnO c-axis, TO modes are allowed. The A_1(TO) mode at about 378 cm^{-1} is observed in $x(yy)\bar{x}$ and $x(zz)\bar{x}$ configuration, while the E_1(TO) mode occurs at about 410 cm^{-1} in $x(yz)\bar{x}$ and $x(zy)\bar{x}$. In $x(zz)\bar{x}$ configuration, the E_2 modes are forbidden and, consequently, do not appear. In contrast, the E_1(LO) mode is clearly observed at about 588 cm^{-1}, although also symmetry-forbidden in $x(zz)\bar{x}$ [Cusco 2007]. However, the E_1(LO) mode was observed before in this configuration, which was attributed to interband Fröhlich interaction in non-ideal ZnO [Calleja 1977].

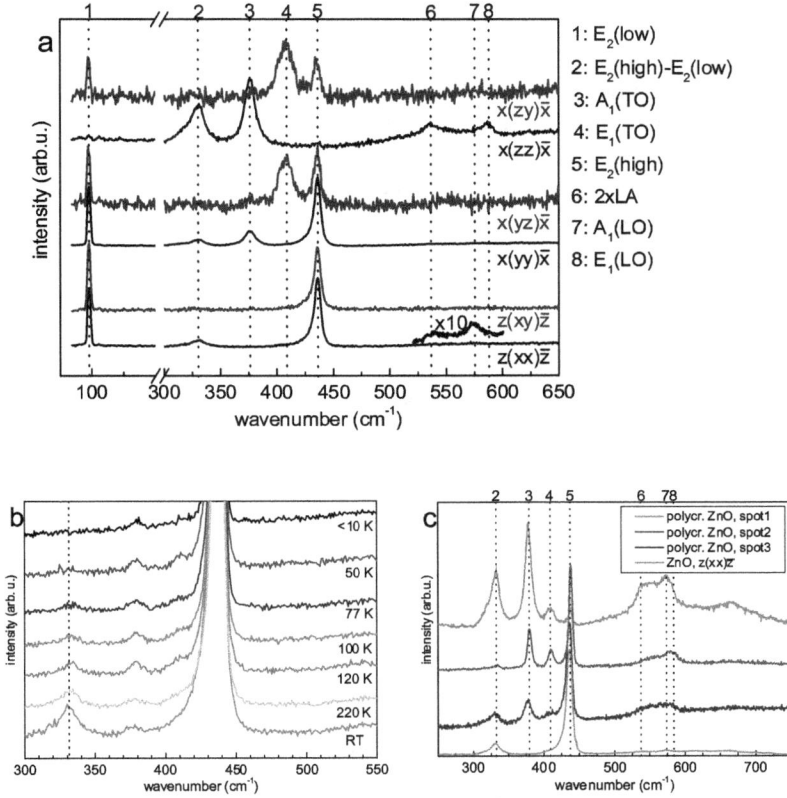

Figure 4.1: *(a) Raman spectra of a ZnO single crystal, recorded in different scattering configurations (excitation: $\lambda = 514.5$ nm). The scattering configurations are denoted using the Porto notation, the mode assignment is in accordance with literature results [Cusco 2007]. (b) Temperature-dependent Raman spectra of a ZnO single crystal (excitation: $\lambda = 514.5$ nm). All spectra are normalized to the intensity of the E_2(high) mode at about 437 cm^{-1}. The mode observed at about 332 cm^{-1} disappears at low temperatures and is assigned to the difference process E_2(high)-E_2(low). (c) Raman spectra of polycrystalline ZnO, recorded with the laser focus on different spots on the surface of the sample (excitation: $\lambda = 514.5$ nm). Because the scattering configuration is not well-defined for such a polycrystalline sample, the spectra reflect the mixture of orientations present within the laser spot during each experiment. For comparison, a Raman spectrum of a ZnO single crystal is shown, recorded in $z(xx)\bar{z}$ configuration.*

In contrast to the predictable Raman spectra in a well-ordered ZnO single crystal, the spectra in Figure 4.1c can not be attributed to one defined symmetry. The micro-Raman experiments for these spectra were conducted with the laser focused on different spots on a polycrystalline ZnO sample grown by vapor phase transport (VPT). Therefore, the spectra reflect the different orientations which are present within the laser spot of about μm size. In addition, a broad signal appears in the spectra of the polycrystalline sample between 500 cm^{-1} and 600 cm^{-1}, peaking at about 570 cm^{-1} to 590 cm^{-1}. In this wavenumber region of high phonon DOS, the LO phonon modes are located. Their intensity is strongly influenced by Fröhlich interaction (see subsection 2.3.2), which itself can be induced by impurities and intrinsic defects [Colwell 1972, Friedrich 2007]. The band is therefore attributed to disorder-induced Raman scattering, resulting in: (a) intensity-enhanced LO phonon modes, especially $A_1(LO)$, due to disorder-induced Fröhlich scattering, and (b) showing through of the high phonon density of states due to phonon contributions from outside the BZ center. Such contributions are expected in disordered systems due to softening of the Raman selection rules, see subsection 2.3.2. Therefore, the broad band reflects the increased disorder of this VPT-grown polycrystalline sample compared to the ZnO single crystals discussed above. The origin of the 'disorder band' will be discussed in more detail in the course of this thesis. For example, it will be studied for extrinsic disorder in Ar-irradiated ZnO single crystals in subsection 4.1.1.

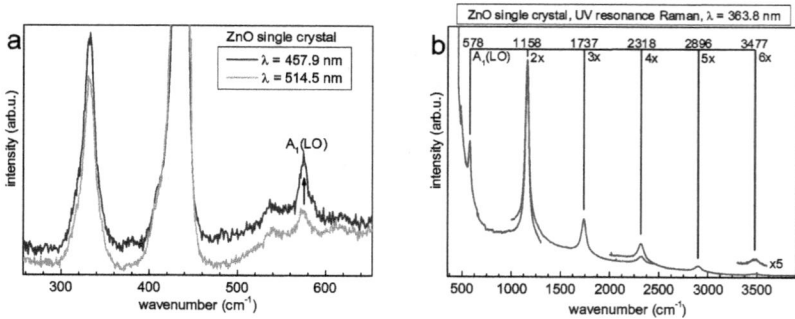

Figure 4.2: *(a) Raman spectra of a ZnO single crystal, recorded with different laser wavelengths (excitation: $\lambda_1 = 514.5$ nm, $\lambda_2 = 457.9$ nm). A weak resonance effect is observed for the $A_1(LO)$ mode at about 577 cm^{-1}. (b) Resonant Raman scattering in a ZnO single crystal (excitation: $\lambda = 363.8$ nm). The spectrum is solely dominated by the $A_1(LO)$ mode and the corresponding multiple phonon processes $2xA_1(LO)$, $3xA_1(LO)$, etc.*

With regard to the experiments of chapters 5 and 6, it can be stated that the CrysTec ZnO single crystals show the expected ZnO Raman scattering fingerprint for all configurations. In particular, no additional modes are observed. Moreover, the peak widths and frequency positions indicate a very good crystalline quality. The VPT-grown sample is polycrystalline and shows indications for a somewhat inferior structural quality. While other VPT-grown samples studied in this thesis show a significantly better crystal quality, their polycrystalline character inevitably leads to less well-defined scattering configurations.

Figure 4.2 shows resonance effects in the Raman scattering results of a ZnO single crystal. First resonance effects are already observed when tuning the wavelength of the exciting laser from green (514.5 nm) to blue (457.9 nm), see Figure 4.2a. The stronger resonance of the LO modes associated with Fröhlich scattering compared to the deformation-potential-dominated modes leads to a higher intensity of the $A_1(LO)$ mode when tuning the laser wavelength towards the ZnO band gap energy [Calleja 1977]. This resonance effect is much stronger when the exciting laser has an energy exactly at or very close to the ZnO band gap at about 3.4 eV. This is shown in Figure 4.2b. The spectrum taken with the 363.8 nm line of an Ar ion laser is solely dominated by the $A_1(LO)$ mode and the corresponding multiple modes $2xA_1(LO)$, $3xA_1(LO)$, etc. Raman resonance experiments provide sensitivity advantages, which are especially useful for very thin layers or other systems which show weak Raman scattering intensity. However, for most of the systems studied in this work, excitation in the visible range delivers more information at a clearly sufficient intensity. Moreover, UV excitation can lead to strong local heating effects in ZnO, especially in nanocrystallites [Alim 2005/1, Alim 2005/2]. For this reason, excitation in the visible spectral range combined with sufficiently long integration time were chosen in order to obtain the required signal strength in the Raman experiments for this thesis.

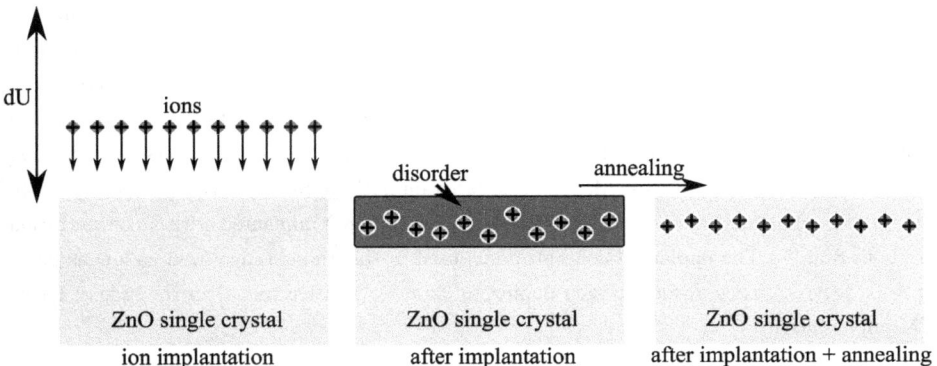

Figure 4.3: *Schematic diagram of an ion implantation process as conducted for several ZnO systems of this thesis. Ions are accelerated electrostatically and impinge on the target crystal. Besides the incorporation of the desired elements, crystal damage is induced by the implantation, which can be healed to a large extent by thermal annealing.*

4.1.1   Effect of ion irradiation on ZnO single crystals

In this section, the commonly used method of ion implantation is outlined. Since pure ZnO is in the focus of this chapter, the experimental analysis is restricted to the disorder induced by the ion irradiation. Additional disorder effects of the implanted ions are discussed in chapter 5 for the incorporation of transition metal ions (magnetic alloying) and in chapter 6 for the incorporation of nitrogen ions (p-type doping). In both cases, the host crystals are hydrothermally grown ZnO single crystals from CrysTec, Berlin, which were characterized by Raman scattering above. The host crystals have a well-defined orientation and are implanted along the (0001) direction using the Zn-polar surface. The implantation process is shown schematically in Figure 4.3. Ions of the desired element are produced within an ion source and electrostatically accelerated. In a target chamber, they impinge on the host crystal. The amount of implanted material is very well tunable via the ion energies and fluences. Furthermore, only the desired ions are incorporated and the implantation of other impurities can be excluded to the greatest extent. Besides the incorporation of the desired element, also disorder is accumulated at the surface and in the implanted layer. In order to obtain samples with good structural quality, thermal annealing is generally required after an ion implantation. Theoretical considerations and experimental results concerning ion-implanted ZnO and subsequent thermal annealing are summarized in a recent review article [Kucheyev 2006].

The penetration depth of the ions depends on the ion energy and on the material properties of both the implanted ions and the host crystal. Energies in the keV range correspond to actual ion implantation, lower energies result in ion beam deposition on the target surface. For the implantations discussed in this thesis, energies between 50 keV and 450 keV as well as total fluences from 1×10^{13} cm^{-2} to 4×10^{17} cm^{-2} were used. For each implantation, a set of fluences and energies was chosen in order to achieve a box-like implantation profile with a constant concentration within a layer of several 100 nm depth. To determine the ideal implantation parameters, implantation profiles were calculated with the Monte Carlo program package SRIM/TRIM [Ziegler 1985]. In Figure 4.4a, the result of such a calculation is shown for ZnO implanted with a concentration of about 8 at.%. The implanted concentrations used in this thesis range from as low as 0.005 at.% (relative to oxygen) for nitrogen doping to 32 at.% (relative to Zn) in the case of heavy TM implantation.

Obviously, ion implantation generally leads to significant crystal disorder throughout the implanted region. Ions lose their kinetic energy via collisions with target atoms and, continuously, via an energy drag from the overlap of electron orbitals. The energy transfer to the host crystal results in point defects like interstitials and vacancies. Using SRIM/TRIM, the ballistic processes are simulated, which lead to atom displacements in the host crystals [Ziegler 1985]. The results

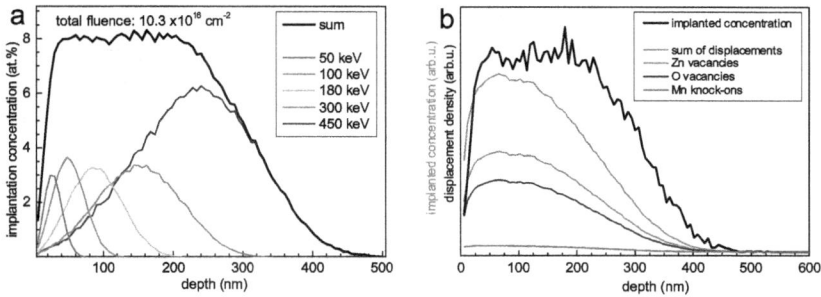

Figure 4.4: *(a) Simulations for ion-implanted ZnO host crystals, calculated with the Monte Carlo program package SRIM/TRIM [Ziegler 1985]. The energies and fluences of the ions have been chosen in order to obtain a box-like implantation profile with a maximum concentration of 8 at.%. (b) Implantation and atom displacement profiles, respectively, of ion-implanted ZnO, calculated with the Monte Carlo program package SRIM/TRIM. The total displacement profile is shifted to a slightly lower depth relative to the implantation profile because the highest damage by an implanted ion is induced shortly before it comes to stop.*

of such a simulation are shown in Figure 4.4b, using the example of implantation with 2 at.% of manganese ions into ZnO. The displacement profile looks similar to the implantation profile, but is shifted to slightly lower depths. Most of the damage caused by an implanted ion, i.e. the largest energy transfer to the crystal, is accumulated shortly before the ion comes to a stop. Still, such Monte-Carlo simulations only compute ballistic processes and neglect the effect of dynamic annealing by the ions during the implantation process. Experimental results suggest that this dynamic annealing is particularly strong for ZnO due to the high ionicity of the Zn-O bond [Kucheyev 2006]. It promotes defect migration and, therefore, ZnO usually stays crystalline, even when implanted with heavy atoms using high doses. Upon such heavy implantation, another effect has to be taken into account: Sputtering can etch away several monolayers of the crystal's surface.

The disorder induced in the host crystal by the implantation depends not only on implantation parameters, like energies and fluences, but also on the implanted element through its chemical properties and mass. The disorder can therefore be divided in (i) irradiation disorder and (ii) disorder induced by the impurities implanted into the host crystal. The disorder effects induced by the implanted impurities are discussed in detail in the chapters 5 and 6 for transition metal and nitrogen implantation, respectively. In this chapter, disorder effects of pure ZnO are in the focus. To analyze the implantation effect on Raman scattering without strong effects of the incorporated impurity ions, ZnO crystals were irradiated with Ar ions. Due to their chemically inert character, they are not expected to induce strong impurity effects, but mostly irradiation damage. A considerable amount of surface-near implanted Ar ions can be expected to leave the crystal already during the implantation or during the subsequent thermal annealing. Thus, the process is referred to as Ar irradiation and the total fluence is used to describe the process and not a concentration value. The total Ar fluences chosen for the characterization of the irradiation damage were 1.6×10^{16} cm^{-2}, 3.1×10^{16} cm^{-2}, 6.3×10^{16} cm^{-2}, and 12.6×10^{16} cm^{-2}. For each sample, a set of energies between 60 keV and 300 keV was chosen for a box-like irradiation profile. The smaller mass implies also that the disorder effect of Ar irradiation is significantly smaller than of implantation with heavier elements such as the TM ions used in chapter 5. Nevertheless, the Ar irradiation provides a good model system to study the effect of such extrinsic disorder.

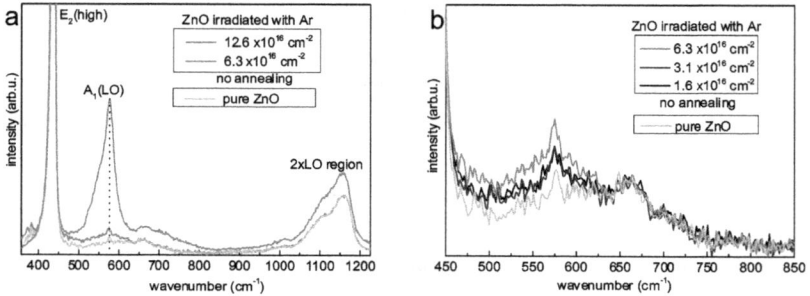

Figure 4.5: *(a) Raman spectra of ZnO single crystals irradiated with Ar, using fluences of 6.3 $x10^{16}$ cm^{-2} and 12.6 $x10^{16}$ cm^{-2} , respectively (excitation: $\lambda = 514.5$ nm). For comparison, a spectrum of pure ZnO is shown. The disorder is reflected by a broad band in the A$_1$(LO) region, especially for the 12.6 $x10^{16}$ cm^{-2} irradiation. (b) Raman spectra of ZnO single crystals irradiated with Ar, using fluences of 1.6 $x10^{16}$ cm^{-2}, 3.1 $x10^{16}$ cm^{-2}, and 6.3 $x10^{16}$ cm^{-2}, respectively (excitation: $\lambda = 514.5$ nm). For comparison, a spectrum of pure ZnO is shown. The disorder effect scales with the irradiation dose, but is relatively weak in these samples irradiated with comparatively low Ar doses.*

Figure 4.5a shows the Raman spectra of ZnO crystals, irradiated with 6.3×10^{16} cm^{-2} and 12.6×10^{16} cm^{-2} Ar, respectively. In addition, a Raman spectrum of pure ZnO is shown for comparison. All spectra exhibit a strong E$_2$(high) mode at about 437 cm^{-1}, as it can be expected in this standard backscattering experiment. The most obvious difference between the spectra of the irradiated crystals and the pure ZnO spectrum occurs between 500 cm^{-1} and 600 cm^{-1}. Especially in the spectrum corresponding to the highest Ar fluence, a broad band appears, peaking near the A$_1$(LO) position at about 576 cm^{-1}. Note that the intensity of this band is still below 20% of the E$_2$(high) mode maximum for this Ar-implanted sample, in contrast to samples with heavier implantation damage discussed in the course of this thesis. Figure 4.5b shows that this effect of increased intensity at the $A_1(LO)$ position is already observable for lower irradiation fluences, but to a strongly reduced degree.

Several points indicate that the broad, additional Raman feature in Figure 4.5a is in fact the intensified $A_1(LO)$ mode, as it was assigned above for the case of disordered, polycrystalline ZnO. For smaller fluences, the spectra of the irradiated samples look nearly identical to the pure ZnO spectrum with only a slight intensity increase of the $A_1(LO)$ mode (Figure 4.5b). This effect is much stronger for larger fluences, but with an additional broadening of the signal. This broadening can be attributed to Raman scattering contributions of the $A_1(LO)$ phonon branch from outside the Brillouin zone center due to disorder-induced reduction of the crystal symmetry. Accordingly, a red-shift of this band is observed for heavily implanted ZnO, e.g. in subsection 5.2.1, corresponding to the dispersion of the $A_1(LO)$ phonon branch near the Γ-point of the Brillouin zone. The higher intensity of the $A_1(LO)$ phonon is additionally reflected in stronger second-order signals in the 2xLO region between 1000 cm^{-1} and 1200 cm^{-1} in Figure 4.5a. Furthermore, the identity of the disorder band with the intensified and broadened $A_1(LO)$ mode is confirmed by Raman experiments using different scattering configurations in subsection 6.1.2: The $A_1(LO)$ is symmetry-forbidden in $x(yy)\bar{x} + x(yz)\bar{x}$ configuration, and, accordingly, the broad disorder band is not observed in the corresponding experiments.

What is the mechanism behind the strong intensification and broadening, observed only for the $A_1(LO)$ mode? The effect is obviously connected to disorder, as it scales with the used irradiation dose. In the UV experiments of this section, it was shown that it is also the $A_1(LO)$, which is affected most strongly by Raman resonance effects in ZnO. In this case, the effect was attributed to the stronger resonance of LO modes, which are dominated by Fröhlich scattering. Fröhlich scattering, on the other hand, can be induced by impurities or defects in a crystal [Colwell 1972, Friedrich 2007]. In summary, the broad disorder band between 500 cm^{-1} and 600 cm^{-1}, which is observed in many experiments throughout this thesis, is attributed to intensification and broadening of the $A_1(LO)$ mode, caused by extrinsic Fröhlich scattering and Raman scattering contributions from outside the Brillouin zone center. The intensity, peak width, and frequency position of this Raman signal can be used as a measure for the crystal quality of ZnO.

Because ion irradiation damages the crystal structure of the target, mostly a subsequent thermal annealing is applied. The thermal energy causes defects to migrate, which can improve the quality of the host crystal and may support substitutional incorporation of the implanted species, in addition [Erhart 2006]. On the other hand, the migrating defects can also form defect clusters or secondary phases containing the implanted ions.

In this section, the healing effect of thermal annealing on Raman spectra is studied, using again the example of Ar-irradiated ZnO single crystals. With thermal annealing, both a recovery of

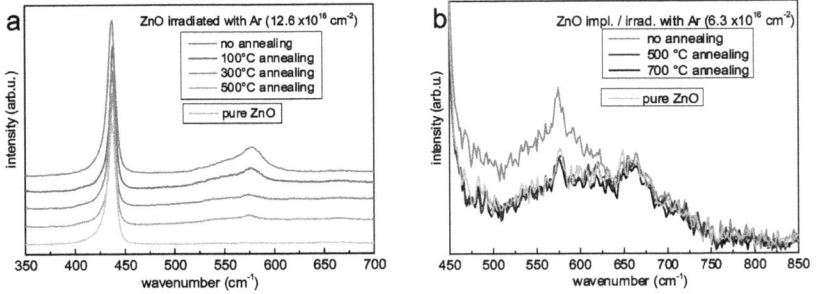

Figure 4.6: *(a) Raman spectra of a ZnO single crystal irradiated with Ar, using an ion fluence of 12.9 $x10^{16}$ cm^{-2} (excitation: λ = 514.5 nm). Thermal annealing with $T_{ann} \geq$ 300 °C results in a substantial healing of the implantation-induced crystal disorder. (b) Raman spectra of a ZnO single crystal irradiated with Ar, using an ion fluence of 6.3 $x10^{16}$ cm^{-2} (excitation: λ = 514.5 nm). For this implantation dose, the crystal disorder is completely healed by thermal annealing at $T_{ann} \geq$ 500 °C within the sensitivity of the conducted Raman experiments.*

the ZnO crystal quality in the implanted layer is achieved and the transparency of the layer is increased, causing more Raman signal to originate from the underlying intact ZnO bulk. Therefore, no attempt is made to evaluate the annealing-induced lattice recovery from the Raman spectra quantitatively. For the Ar-irradiated samples, post-implantation annealing was performed for 15 to 30 minutes at various temperatures from 100 °C to 700 °C in air. In Figure 4.6a, the Raman spectra of Ar-irradiated ZnO (12.6 $x10^{16}$ cm^{-2}) are shown after different annealing steps. It can be seen that annealing at 100 °C for 15 min has no substantial healing effect. But already annealing at 300 °C for 15 min heals most of the Ar irradiation damage, reflected in an intensity reduction of the disorder band peaking at about 576 cm^{-1}. For lower Ar doses, already thermal annealing with 500 °C for 15 min completely heals the crystal disorder within the sensitivity of the applied Raman method, as shown in Figure 4.6b. The spectrum of the 500 °C annealed sample and the pure ZnO sample concur and no further improvement is observed for an additional 700 °C annealing step of 30 min. Normally, thermal annealing at approximately 2/3 of the melting temperature (T_{melt} = 1975 °C for ZnO) is required to heal the complete damage in semiconductors after heavy ion bombardment [Kucheyev 2001, Kucheyev 2006]. Here, substantial healing occurs already for significantly lower temperatures, starting from 300 °C. One probable reason is the reported damage-induced reduction of the thermal stability specific to ZnO [Kucheyev 2006].

The disorder effects in the implanted systems of chapters 5 and 6 are generally stronger than the effect observed for Ar irradiation. In particular, the implanted ions induce strong impurity effects: Additional Raman modes by secondary phase formation and by impurity-activated silent modes are observed for TM-alloyed and nitrogen-doped ZnO crystals, respectively. Furthermore, the well-defined orientation of the ZnO host crystals can be locally distorted upon heavy implantation and symmetry-forbidden TO phonon modes can occur. The required annealing temperature depends not only on the disorder induced by the irradiation, i.e. on ion mass, dose, and energy, but also on the chemical properties of the respective elements [Chen 2006]. For high implantation doses, secondary phase segregation is induced by thermal annealing in TM-implanted ZnO, see chapter 5. As will be shown, the formation of oxide secondary phases is favored by annealing in air.

4.2   ZnO nanoparticles

The ZnO nanostructures studied for this thesis show several size-related Raman scattering characteristics. Some of them can be observed in the Raman spectra of wet-chemically synthesized nanorods deposited on ZnO substrate, as shown in Figure 4.7. Because the deposited nanorods have random orientation, no defined scattering configuration can be achieved. Therefore, the observed Raman modes with high intensity include the E_2(high)-E_2(low) at about 330 cm^{-1}, the A_1(TO) at about 378 cm^{-1}, the E_1(TO) at about 410 cm^{-1}, and the E_2(high) at about 437 cm^{-1}. A mode with small intensity occurs in the LO phonon region at about 581 cm^{-1}. This frequency is between the above observed bulk values of the A_1(LO) mode at about 577 cm^{-1} and the E_1(LO) mode at about 588 cm^{-1}. The feature can be attributed to a quasi-LO mode, which is a mixed symmetry mode due to phonons propagating between the a-axis and the c-axis [Bergman 1999, Loudon 1964], see subsection 3.1.2. If a nanocrystalline ensemble exhibits a preferential orientation, this is reflected in the frequency value of this quasi-LO mode [Bergman 2005]. For a pure random orientation of ZnO crystallites, Bergman et al. calculated the frequency of the quasi-LO mode as 580 cm^{-1}, which is in good agreement with the observed value in the nanorod spectrum of about 581 cm^{-1}. The growth of such nano-sized crystallites usually leads to a reduced structural quality. Consequently, the quasi-LO is intensified in the Raman spectrum by the same mechanism as discussed for the A_1(LO) mode in disordered ZnO in section 4.1. This effect is not strong for the studied ZnO nanorods, reflecting a comparatively good crystalline quality.

Additional Raman modes are observed at about 750 cm^{-1} to 800 cm^{-1} and at about 970 cm^{-1} in Figure 4.7. These modes correspond to vibrations of organic molecules, which are generally used as the stabilizers of nanocrystals grown in wet-chemical syntheses. Such organic ligands

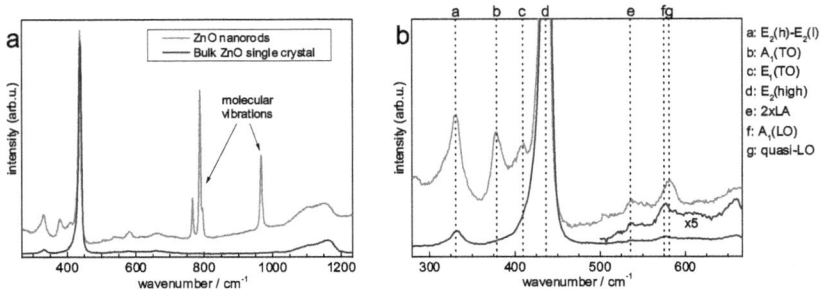

Figure 4.7: *(a) Raman spectra of wet-chemically synthesized ZnO nanorods (excitation: $\lambda = 514.5$ nm). Besides the ZnO phonon modes, also molecular vibrations of the organic ligands are observed. (b) Raman spectra of (a) shown in the low-frequency region. The Raman feature at about $581\ cm^{-1}$ is attributed to the quasi-LO mode reflecting the random orientation of the nanorods.*

are much more sensitive to temperature effects than the ZnO crystallites. In addition, ZnO nanocrystals exhibit a strongly reduced heat conduction compared to bulk material. Therefore, laser-induced temperature effects play an important role in Raman scattering experiments of ZnO nanosystems with organic ligands. Moreover, also the ZnO phonons are affected by local heating due to thermal expansion and anharmonic phonon coupling [Alim 2005/1, Alim 2005/2]. Such local heating effects are especially strong in micro-Raman scattering experiments, as conducted for this thesis, because the focused laser spot leads to an increased laser power density. As a direct size effect, optical phonon confinement was reported for nanostructured ZnO [Fonoberov 2004, Rajalakshmi 2000]. No confinement effects are observed in the Raman spectra of the ZnO nanorods in Figure 4.7 compared to the bulk values. However, the average crystallite size of this nanorod ensemble was not analyzed and, therefore, the confinement effect is studied in more detail on wet-chemically synthesized ZnO nanoparticles with well-defined size and structural properties.

The ZnO nanoparticles were grown using a newly developed wet-chemical synthesis procedure [Chory 2007]. Besides controlled particle size and optimized structural properties of the ZnO nanoparticles, the objective of this synthesis was to obtain large amounts of pure, nanocrystalline powder.

In the following, the used synthesis is shortly outlined using the example of acetate-stabilized ZnO nanoparticles. As precursor material, zinc acetate dihydrate $Zn(OAc)_2$ is dissolved in ethanol and stirred. Step by step, the base tetramethylammonium hydroxide (TMAH) is added in order to bind by-products of the ZnO cluster formation. Finally, the nanoparticles are precipitated by the addition of hexane, centrifuged, and dried in a desiccator. For the growth of ZnO nanoparticles capped with another organic ligand than acetate, the synthesis is modified by adding the corresponding ligand material before the addition of TMAH. Further synthesis details are provided in [Chory 2007].

In a first synthesis series, different growth parameters and stabilizing molecules were analyzed [Chory 2007]. Detailed structural investigations of such nanoparticle systems based on X-ray powder data are reported, using different approaches: via the Debye equation [Kumpf 2005], via the pair distribution function [Neder 2005], and by an explicit modeling of the nanoparticles [Niederdraenk 2007]. For the nanoparticles studied in this thesis, characterization by these XRD methods revealed high stacking fault densities and a size-dependent anisotropic shape [Chory 2007, Neder 2007, Niederdraenk 2007]. The experimental details of the XRD measurements mentioned in this section can be found in [Chory 2007, Neder 2007].

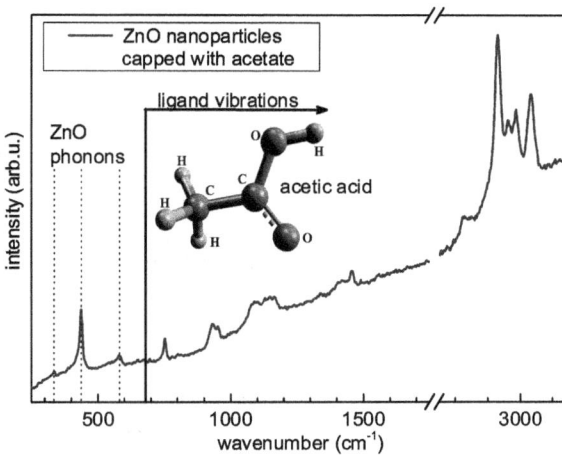

Figure 4.8: *Raman spectrum of acetate-capped ZnO nanoparticles with an average diameter of about 12 nm (excitation: $\lambda = 457.9$ nm). ZnO phonon modes as well as molecular vibrations of the organic ligands are observed. The inset shows the structural formula of the added stabilizer material.*

Figure 4.8 shows the Raman spectrum of acetate-capped ZnO nanoparticles, which were grown according to the synthesis described above. XRD analysis revealed an average particle size of about 12 nm and an ellipsoidal shape for these crystallites [Chory 2007]. In the lower-wavenumber region, characteristic ZnO phonon modes are observed: E_2(high)-E_2(low) at about

333 cm^{-1}, E$_2$(high) at about 437 cm^{-1}, and quasi-LO at about 580 cm^{-1}. The vibrational frequency of the quasi-LO mode corresponds to a random orientation of the nanoparticles [Bergman 2005]. Its rather strong intensity indicates a reduced crystal quality, which is in accordance with the high stacking fault density deduced from XRD. However, the intensity of the quasi-LO mode is intensified by resonance effects in addition, as the spectrum was recorded with the blue 457.9 nm line of the Ar ion laser. The positions of the E$_2$(high) mode and the E$_2$(high)-E$_2$(low) mode agree well with the values of bulk ZnO in section 4.1. No peak shifting due to size effects is observed. In addition to the ZnO phonon modes, molecular vibrations of the organic ligands occur between about 750 cm^{-1} and 3100 cm^{-1}. Important organic vibrations are C-C stretching and C-H bending modes, with frequencies between about 1400 cm^{-1} and 1500 cm^{-1}, and C-H stretching modes between 2800 cm^{-1} and 3100 cm^{-1}. By comparison of the observed molecular vibrations with the vibrations of the pure ligand molecules, not all vibrations can be attributed to the acetate ligand, but a mixture of different molecules is required to account for the observed Raman features. In particular, vibrations of TMAH are observed in many spectra of the nanoparticles grown with this synthesis, independent from the used stabilizing molecules [Raskin 2008]. While this does not necessarily affect the quality of the nanoparticle cores, i.e. the ZnO crystallites, this finding is important for systems where the ligand properties are crucial. This is the case, for example, if the toxicity of the particles plays a major role or if nanoparticles shall be grown which are solely capped with specific functional ligands. An example for such a case is discussed in the following.

To obtain nanoparticles capped with functional ligands, the synthesis described above was modified by the addition of the dye molecule oracet blue as stabilizer before TMAH was added. The structural formula of oracet blue is shown as inset in Figure 4.9a. The Raman spectra of this sample show no vibrations which could be attributed to TMAH molecules and, in addition, FT-NIR Raman experiments indicate the attachment of oracet blue molecules as major ligands [Schumm 2005]. Comparable dye-capped nanoparticles are already used as markers in biology and medicine. For such applications, it is crucial that the particles are stabilized by the intended ligands, and not by a mixture of several synthesis by-products. Figure 4.9a shows two Raman spectra of the oracet-blue-capped ZnO nanoparticles. These spectra are the first and the last spectrum of a series of consecutively registered Raman spectra during an experiment with varying laser power. This experiment was conducted with the 457 nm laser line of the Ar ion laser. To study the contribution of local heating, the laser power density was gradually tuned from 600 kW cm^{-2} at the beginning of the experiment to 300 kW cm^{-2} at the end of the experiment over 24 minutes. The Raman spectrum taken at the beginning of the experiments shows a very intense quasi-LO mode corresponding to strong structural disorder in the nanocrystalline core of the particles. In contrast, the intensity of the quasi-LO mode is strongly reduced at the end

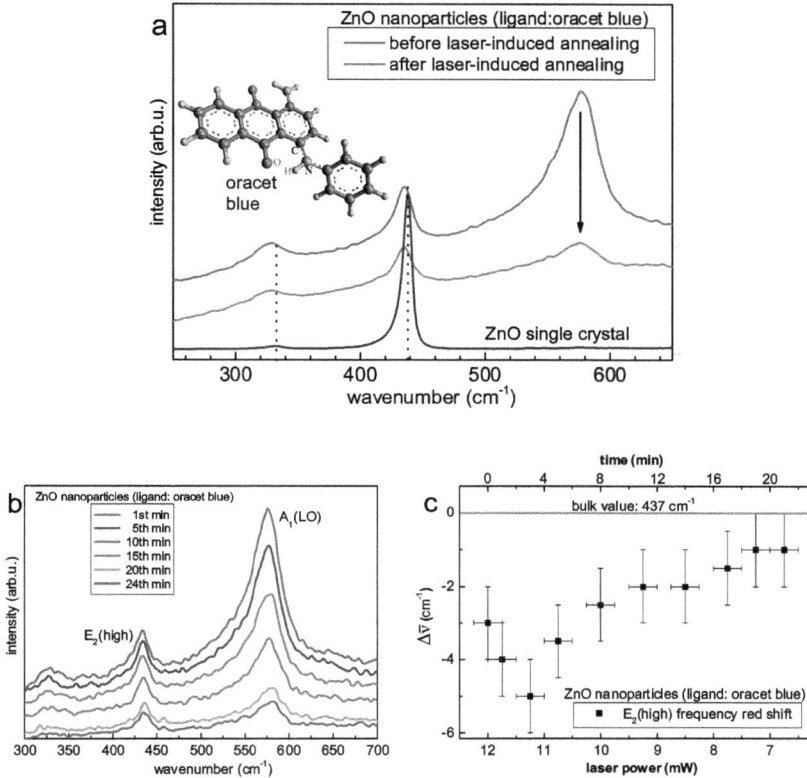

Figure 4.9: *(a) Raman spectra of ZnO nanoparticles capped with oracet blue ligands (d ≈ 12 nm), recorded at the beginning and at the end of a laser-induced annealing experiment (excitation: λ = 457.9 nm). For comparison, the spectrum of a bulk ZnO single crystal is shown. The structural formula of the stabilizer oracet blue is shown as inset. (b) Raman spectra of the same nanoparticle sample, taken during the experiment of (a) (excitation: λ = 457.9 nm). (c) Corresponding red shift of the ZnO E_2(high) mode during the experiment.*

of the experiment, what indicates a strongly enhanced structural quality. This healing effect is attributed to laser-induced annealing of the particles. By Raman spectra taken at the same spot of the powder sample, but several minutes after the experiments, the irreversibility of this effect was confirmed. Due to their small size, such nanoparticle ensembles show reduced heat conduction and, in consequence, the laser power densities of up to 600 kW cm^{-2} induce high temperatures within the laser spot region. Comparable power densities led to temperatures of about 700 °C in Raman experiments on nanoparticles of comparable size [Alim 2005/1, Alim 2005/2]. However, Alim et al. used UV excitation in their experiments, which can be expected to induce very much stronger local heating than the 457.9 nm excitation in the transparent regime, which was used in the above described experiment. Nevertheless, the local temperature is in a range which was successfully applied for thermal annealing of bulk ZnO in section 4.1.

Figure 4.9b shows Raman spectra taken during this experiment. The annealing effect is strongest in the first half of the experiment, i.e. at higher laser power density. Besides the intensity ratio quasi-LO/E_2(high), also the vibrational frequencies of the phonon modes change during the experiment. Figure 4.9c shows the red shift of the E_2(high) mode compared to the bulk value during the same laser-induced annealing experiment. In the first three minutes, the red shift increases from about 3 cm^{-1} to about 5 cm^{-1}. This is attributed to the heating up of the particles at the beginning of the experiment. Consequently, the reduction of the red shift from about 5 cm^{-1} to about 1 cm^{-1} towards the end of the experiment is attributed to the reduced local heating of the sample due to the decreasing laser power. Additionally, the structural improvement by the laser-induced annealing could contribute to a reduction of the red shift. However, the clearly observed heating effect at the beginning of the experiments indicates that temperature effects dominate the E_2(high) peak shifts. This is in accordance with temperature effects reported for micro-Raman experiments on ZnO nanoparticles in the literature [Alim 2005/1, Alim 2005/2]. It should be noted that also ligand effects are expected to play a major role in such experiments. At temperatures of up to 700 °C, most organic ligand molecules are degraded, leading to proceeding clustering of the nanocrystallites.

For the oracet-blue-capped ZnO nanoparticles (d $\approx$ 12 nm), all shifts were explained by local heating and structural effects. However, optical phonon confinement was reported for ZnO nanoparticles between 4 nm and 8 nm [Fonoberov 2004, Rajalakshmi 2000].

Figure 4.10 shows the Raman spectra of ZnO nanoparticles with different organic ligands and average diameters between about 2 nm and 16 nm. The spectra of the nanoparticle samples with average diameters of about 4.0 nm, 6.2 nm, and 16 nm were recorded using the 457.9 nm laser line of the Ar ion laser. The experiments were conducted at low power densities <100 kW cm^{-2} and with very long integration time in order to avoid temperature effects due to local heating. These larger particles were grown in the first synthesis series and characterized in [Chory 2007]. The samples with very small average diameters of about 2.0 nm, 3.2 nm, and 4.7 nm were grown in a second series. Their growth was still according to the synthesis described above, but after optimization of growth parameters like time intervals and relative concentrations of the used chemicals with regard to a further particle size reduction [Pfeiffer 2007]. Furthermore, XRD experiments were conducted to determine structural properties and the average size of

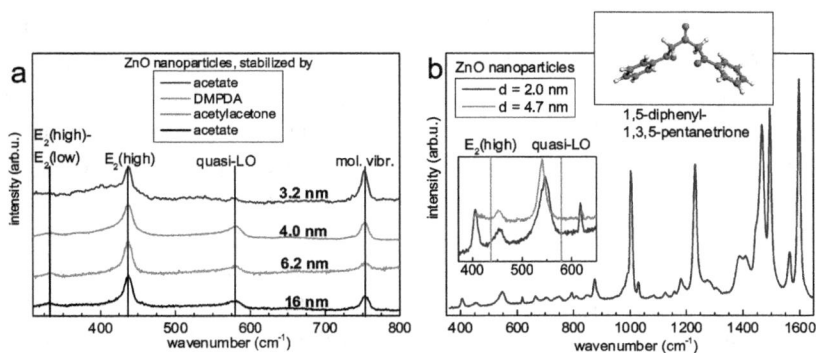

Figure 4.10: *(a) Raman spectra of ZnO nanoparticles with average diameters between about 3.2 nm and 16 nm, capped with various ligands (excitation: $\lambda = 632.8$ for the nanoparticles with $d = 3.2$ nm, $\lambda = 457.9$ nm for the other samples). (b) Raman spectra of ZnO nanoparticles with pentanetrione as stabilizing ligand and average diameters of about 2.0 nm and 4.7 nm (excitation: $\lambda = 632.8$ nm). The stabilizing molecule is shown as inset. No ZnO phonon modes are observed for these nanoparticle samples.*

the particles [Pfeiffer 2007]. The Raman spectra of these nanoparticle samples were recorded at an even lower power density <50 kW cm^{-2} with the 632.8 nm laser line of a He-Ne laser. Because of this excitation with a laser wavelength in the red spectral region, the quasi-LO mode is comparatively weak, but it is still observable for the nanoparticle samples with an average diameter of about 3.2 nm.

Obviously, the ZnO phonon modes have the same vibrational frequencies for all nanoparticle samples shown in Figure 4.10a. In particular, no influence of the average size of the crystallites or of the stabilizing ligand material on the ZnO modes is observed. The Raman feature at about 750 cm^{-1} is attributed to TMAH molecular vibrations. Therefore it can be stated that TMAH is present as part of the stabilizing organic material for all particle samples shown in Figure 4.10a. No ZnO phonon modes could be observed for the even smaller nanoparticles with an average diameter of about 2 nm in Figure 4.10b, despite the clear signature of ZnO nanocrystallites in XRD experiments. For their growth, the above described synthesis was modified as follows. The precursor Zn(OAc)$_2$ was dissolved in a ethanol-hexane mixture and the ligand material 1,5-diphenyl-1,3,5-pentanetrione, referred to as pentanetrione in the following, is added before the addition of TMAH [Pfeiffer 2007]. With their average diameter of only about 2.0 nm, these samples are among the smallest ZnO nanocrystallites grown until now.

Several reasons for the absence of ZnO phonon modes in the Raman spectra of the smallest nanoparticles are conceivable: (i) structural quality: Using a modified Rietveld method, the smallest nanoparticles were found to show a very high stacking fault density of nearly 20% [Pfeiffer 2007]. Such a reduced crystal quality could strongly affect the intensity of the Raman scattering signal. (ii) Chemical environment: The nanoparticles were studied by a dynamic light scattering (DLS) method, which indicates that the crystallites are incorporated in an organic matrix [Raskin 2008]. The Raman spectra of the pentanetrione-capped nanoparticles show no TMAH vibrations and the observed molecular vibrations suggest pentanetrione as major material within this organic matrix [Raskin 2008]. The quantity ratio of nanocrystalline ZnO material relative to pentanetrione could not be determined. If the organic ligands should dominate, the Raman signals of the ZnO phonons could be too weak to be observed against the background of the molecular vibrations. Furthermore, also the optical properties of the surrounding ligands could affect the Raman scattering signal from the ZnO nanocrystallites. (iii) Optical phonon confinement: This effect was reported in the size range between 4 nm and 8 nm [Fonoberov 2004, Rajalakshmi 2000]. However, a size series (2.0 nm $\leq$ d ≤ 5.5 nm) of ZnO nanoparticles capped with pentanetrione revealed no ZnO phonons in the Raman spectrum of any sample, independent from the average diameter. For example, the spectrum of the sample with an average diameter of about 4.7 nm is shown in the inset of Figure 4.10b. In contrast to the Raman experiments, the XRD signature of ZnO nanocrystallites was clearly observed.

In conclusion, no frequency shifts due to optical phonon confinement are observed for the studied nanoparticle ensembles, even with average diameters down to 3.2 nm, despite the literature reports of optical phonon confinement effects between 4 nm and 8 nm [Fonoberov 2004, Rajalakshmi 2000]. Nanoparticles capped with pentanetrione were grown with even smaller diameters down to 2 nm. However, no ZnO phonon modes could be observed in the corresponding Raman experiments. Their absence is presumably due to the optical properties of the surrounding organics or caused by the reduced structural quality of these very small nanocrystallites.

Chapter 5

Transition-metal-alloyed ZnO

In subsection 3.2.2, theoretical considerations and the experimental situation regarding ZnO alloyed with transition metals (TM) were presented. The objective of such magnetic alloying are ZnO-based diluted magnetic semiconductors (DMS), i.e. ZnO systems with intrinsic ferromagnetic properties due to substitutional incorporation of TM ions on Zn sites. For these systems, carrier-mediated ferromagnetic interaction of the TM ions is predicted. A key question is whether intrinsic, carrier-mediated coupling of the TM ions is responsible for experimentally observed ferromagnetism of ZnO-based DMS, or if the magnetic properties originate from magnetic secondary phases.

The existence and type of free carriers in ZnO are closely related to its crystal quality (subsection 3.2.1). Intrinsic defects due to the TM incorporation may further increase the n-type character of a ZnO host crystal. While there are both theoretical studies requiring n-type and p-type host ZnO, substitutional TM incorporation on Zn sites and a high crystal quality are required in any case. Hence, understanding the exact incorporation behavior of the TM ions and their impact on the ZnO wurtzite crystal structure is crucial for the desired magnetic properties. The objective of the experiments presented in this chapter is to get information about the position of the TM ions within the ZnO host lattice, about the possible formation of TM-related precipitates, and about the impact of the impurity incorporation on the ZnO crystal quality.

Raman scattering proves to be an excellent method for such investigations. Decreased crystal quality of ZnO is reflected in its Raman spectra by peak broadening, peak shifts, and by the relaxation of symmetry selection rules. Substitutional TM incorporation, on the other hand, may be detectable via specific impurity vibrations. In addition, Raman spectroscopy offers a high potential for detecting secondary phases by their characteristic vibrational eigenmodes. Furthermore, using the micro-Raman technique, the local distribution of such secondary phases can be studied by lateral mapping with μm resolution over the surface. Thus, very small and localized segregations can be observed (subsection 2.2.2). Such precipitates may be invisible for other methods (e.g. conventional XRD), but still can have a significant impact on the magnetic properties of such systems. This is not sufficiently considered in many ZnO:TM-related publications. Often, the magnetic properties of these systems are ascribed to a successful realization of a ZnO-based DMS. In many studies, however, the exclusion of secondary phases is solely based on conventional XRD results or on other methods which are not able to detect small and only local precipitates. The sensitivity advantages of the Raman method can be attributed to its local probe character in contrast to the long-range correlation requirement of interference based methods. This advantage was confirmed in a direct comparison of Raman scattering and XRD on Co-alloyed ZnO [Wang 2007]. As a further example, ferromagnetism in Cu-alloyed ZnO was identified as extrinsic due to CuO inclusions detected by Raman scattering [Sudakar 2007]. In contrast, they were invisible for XRD and HRTEM (high-resolution transmission electron microscopy) at concentrations <3%. The basics of Raman spectroscopy and the experimental Raman setups used for this chapter are presented in chapter 2 and Raman scattering on ZnO is discussed in detail in subsection 3.1.2.

Despite the advantages of Raman spectroscopy for the structural study of transition metal incorporation in ZnO, further experimental methods are clearly required to complement the Raman scattering results. For this thesis, additional experiments included for example XRD and HRTEM. For some samples, direct access to magnetic properties was provided by MOKE (magneto-optical Kerr effect), SQUID (super-conducting quantum interference device), and EPR (electron paramagnetic resonance) measurements. Such magnetic results are not in the focus of this thesis, but are still included in the discussion. Indirect access to the magnetic properties was possible by the identification of secondary phases with well-known magnetic properties, especially elemental TM clusters and TM oxides. Experimental details for the complementing methods are provided together with the corresponding results.

The TM-alloyed ZnO samples presented in this chapter were fabricated by vapor phase transport (Mn, Co), ion implantation of hydrothermally grown crystals (V, Mn, Fe, Co, Ni), molecular beam epitaxy (Co), dip-coating (Co), and wet-chemical synthesis (Mn). They possess varying TM concentrations between 0.2 at.% and 32 at.% relative to Zn. None of the studied samples were doped in addition to the magnetic alloying. Hence, they are expected to be n-type. While for such systems the theory by Dietl et al. is not applicable [Dietl 2000], ferromagnetism may be possible according to the works of Sato et al. and Coey et al. [Coey 2005, Sato 2001, Sato 2002], see subsection 3.2.2. Among the different fabrication processes, the series of TM-ion-implanted ZnO comes with several advantages. Within this thesis, it provides the highest variety of different TM elements. Furthermore, the TM concentration is controllable, and the orientation of the hydrothermally grown ZnO host crystals is well-defined ((0001)-face). On the other hand, the irradiation damage caused by the ion implantation must be healed by subsequent thermal treatment. The effect of ion implantation on ZnO was already presented in section 4.1.1, using the example of Ar irradiation. Consequently, in section 5.1 only the implantation and thermal healing effects characteristic for the studied TM ions are presented.

Among the sections 5.2 to 5.6, the results are arranged by the TM elements Mn, Co, Fe, Ni, and V, respectively. Nanostructured systems were studied for Mn- and Co-alloyed ZnO, see subsections 5.2.3 and 5.3.3. Section 5.7 finally concludes the results on the TM-alloyed ZnO systems.

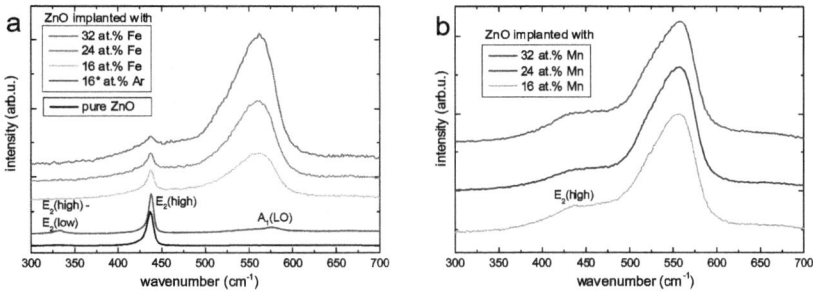

Figure 5.1: *In the Raman spectra of (a) Ar- and Fe-, and (b) Mn-implanted ZnO crystals, the implantation damage is reflected, depending on the implantation concentration and on the implanted material (excitation: $\lambda = 514.5$ nm). The spectra were recorded before any thermal treatment was applied.*

5.1 Effect of transition metal implantation on ZnO

The irradiation effect of ion implantation on a ZnO host crystal was already discussed in section 4.1.1 for Ar ions. In this section, the implantation of various TM ions is studied using micro-Raman spectroscopy. Commercially available ZnO single crystals (CrysTec, Berlin) were implanted at room temperature with V, Mn, Fe, Co, and Ni ions, yielding TM concentrations between 0.2 at.% and 32 at.% relative to Zn. The focus here lies on the samples with concentrations ≥ 8 at.% and especially high implantation damage, as reported in [Schumm 2008/2]. For each implantation process, a combination of up to 5 different ion energies (50-450 keV) was chosen in order to achieve a box-like implantation profile with a resulting layer thickness of about 300 nm [Ziegler 1985]. For comparison, a reference sample was prepared by irradiation of Ar to study the irradiation damage without the additional effect of residual TM impurities. For this irradiation, the energies and the fluence of 16 at.% Fe-implanted ZnO were taken and its concentration is labeled with 16* at.%. Note that this Ar sample corresponds to the Ar-irradiated ZnO crystal with a fluence of 12.6 x10^{16} cm^{-2} in subsection 4.1.1.

There are several Raman spectroscopic studies on implanted ZnO in the literature, e.g. [Chen 2005, Chen 2006, Jeong 2004, Reuss 2004]. Only few works deal with TM-implanted ZnO, most of them with Mn implantation [Mofor 2006, Schumm 2008/1, Venkataraj 2007, Zhong 2006]. In contrast, there are no Raman studies reported so far on Fe-, Co-, or Ni-implanted ZnO.

In the Raman spectra of the TM-implanted samples (e.g. Fe and Mn in Figure 5.1), a very broad and intensive signal in the $A_1(LO)$ mode region occurs, in contrast to the pure ZnO. This signal corresponds to a disorder-intensified and -broadened $A_1(LO)$ mode as identified in section 4.1.1. The mass of the implanted ions plays the expected major role, as seen by a comparison between Fe-implanted and Ar-irradiated ZnO in Figure 5.1a. There is less damage caused by the irradiation of the lighter and chemically inert Ar ions. Ni and Co implantations show a similar impact on the Raman spectra as in the case of Fe, what can be expected due to the similar mass of the TM ions. In contrast, an especially large effect can be observed for Mn-implanted ZnO in Figure 5.1b: The $E_2(high)$ mode is not clearly defined, but only visible as a shoulder of the dominating $A_1(LO)$ disorder band. The different implantation effect on the Raman spectra in the case of Mn compared to the other TM ions can not be due to its mass. The micro-Raman setups used for this thesis have depth of field values from 5 μm to 10 μm. Thus, for pure and therefore transparent ZnO, one accumulates the signal from some μm depth. Typically, in the implanted case with an implantation depth of some 100 nm, Raman signal is detected both from the implanted layer and the intact crystal below. In the case of more heavily implanted ZnO, the implanted layers show a strongly reduced transparency, dependent on the implantation material. In such a case, the Raman signal originates mainly or solely from the implanted layer. It was found that the light absorption of Mn-alloyed ZnO is considerably higher than in other TM-alloyed ZnO systems [Jin 2000, Polyakov 2003]. Therefore, the strong occurrence of the disorder band reflects the higher absorption of the Mn-implanted layer.

On the other hand, the appearance of the broad $A_1(LO)$ signal, even for all 32 at.% TM-implanted ZnO samples, indicates that the ZnO crystal structure is strongly damaged, but still retains its wurtzite lattice character despite the high implantation dose. This can be attributed to the strong influence of dynamic annealing in ZnO (see section 4.1.1).

Thermal treatment was used to improve the crystalline quality of the TM-implanted systems and to support substitutional incorporation of the implanted species on the appropriate lattice sites. Here, the commonly used method of annealing in air is applied. Note that this process, in contrast to vacuum annealing, favors the formation of TM oxide secondary phases [Thakur 2007]. With thermal annealing, not only a recovery of the ZnO crystal quality in the implanted layer is achieved, but also the transparency of the layer is increased. This causes more Raman signal to originate from the underlying intact ZnO bulk. Thus, no attempt will be made for a quantitative evaluation of the annealing-induced lattice recovery from the Raman spectra.

ZnO impl. with	100 °C	300 °C	500 °C	700 °C
Ar	15 min	15 min	15 min	30 min
Mn	15 min	15 min	15 min	15 min
Fe	15 min	15 min	15 min	30 min
Co	-	-	-	30 min
Ni	-	-	15 min	30 min

Table 5.1: *Applied annealing steps for the TM-implanted samples with TM concentrations ≥ 8 at.%.*

Post-implantation annealing was performed at various temperatures from 100 °C to 700 °C in air for 15 to 30 minutes. The exact annealing steps are displayed in Table 5.1. In Figure 5.2, the Raman spectra of Ar-irradiated as well as Mn- and Fe-implanted ZnO crystals are shown after different annealing steps. As discussed in section 4.1.1, annealing at 100 °C has no substantial healing effect, but already annealing at 300 °C heals most of the damage in the case of Ar, see Figure 5.2a. For the TM ions Fe, Co, and Ni, a healing effect beginning from 300 °C was observed for all concentrations. With 500 °C annealing, the damage was further decreased, but still not completely healed. In the case of the 32 at.% Mn-implanted sample in Figure 5.2b, no healing effect was visible in the Raman spectra at all due to the high absorption of the implanted layer. Hence, thermal treatment at higher temperatures was found to be necessary.

There are several studies which try to determine the optimal annealing temperature for TM-implanted ZnO. For example, an almost perfect substitutional incorporation of Fe at Zn sites after vacuum annealing at 800 °C was detected in emission channeling experiments [Rita 2004]. For higher temperatures of 900 °C to 1000 °C, the fraction of Fe ions on Zn sites was considerably reduced due to thermally induced surface defects, stronger Fe diffusion, and Fe cluster formation [Rita 2004]. In a study of Mn-implanted ZnO, annealing of Zn interstitials and other implantation-produced defects was optimal between 800 °C and 1000 °C (in air), but also strong diffusion of the implanted Mn ions was observed between 700 °C and 900 °C [Sonder 1988]. Therefore, besides the healing effect, also precipitation of a secondary phase involving the implanted TM ions may be supported by thermal treatment. A moderate maximum temperature of 700 °C is chosen in the following sections for annealing of the TM-implanted samples with concentrations ≥ 8 at.%. At this temperature, further healing of the implantation damage is achieved, but one must also check for the onset of secondary phase formation [Schumm 2008/2].

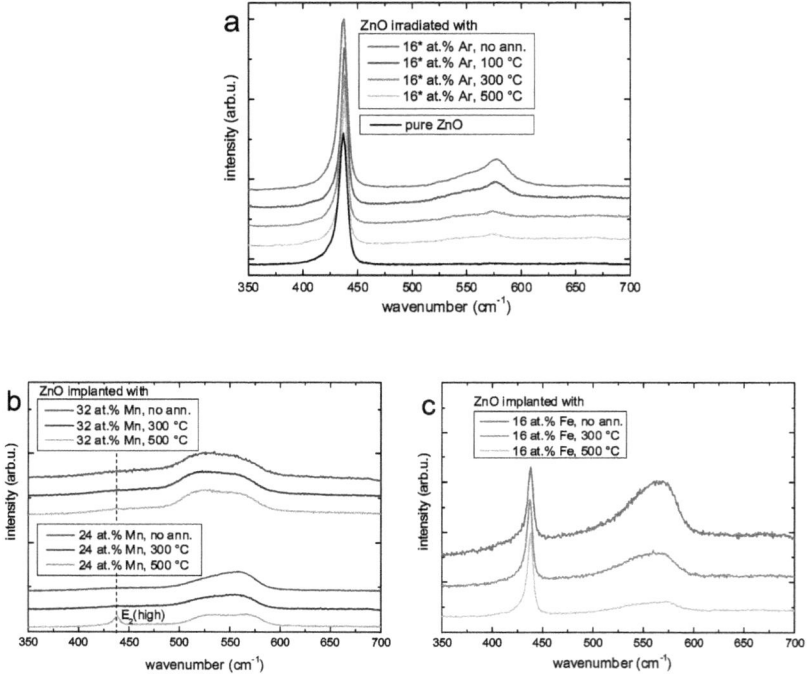

Figure 5.2: *The effect of thermal annealing on the Raman spectra of Ar-irradiated as well as Fe- and Mn-implanted ZnO for temperatures up to 500 °C: (a) pure ZnO and 16* at.% Ar-irradiated ZnO, see also Figure 4.6, (b) 24 at.% and 32 at.% Mn-implanted ZnO, (c) 16 at.% Fe-implanted ZnO (excitation: $\lambda = 514.5$ nm).*

5.2 Manganese-alloyed ZnO

For this thesis, hydrothermally grown ZnO host crystals were implanted with Mn to obtain $Zn_{1-x}Mn_xO$ with 0.2 at.% $\leq$ x $\leq$ 32 at.%. For implantation details, see subsection 4.1.1 and section 5.1. This Mn-implanted series is complemented by polycrystalline bulk ZnMnO, fabricated via a vapor phase transport (VPT) technique, and by wet-chemically synthesized ZnMnO nanoparticles. Table 5.2 gives an overview of all studied ZnMnO samples. Subsection 5.2.1 focuses on the structural impact of Mn impurity incorporation with concentrations $\leq$8 at.%, while for increasing concentrations, secondary phase formation becomes more and more important, see subsection 5.2.2. In subsection 5.2.3, finally, results on Mn-alloyed ZnO nanoparticles are presented.

Mn-alloyed ZnO	Fabrication	Mn concentration
layers	hydrothermally grown ZnO impl. with Mn	0.2-32 at.%
polycrys. bulk	vapor phase transport	< 4 at.%
nanoparticles	wet-chemical synthesis	< 2 at.%

Table 5.2: *Overview of the Mn-alloyed ZnO samples presented in this thesis.*

5.2.1 $Zn_{1-x}Mn_xO$ bulk and layers with concentrations ≤ 8 at.%

There are numerous reports on Raman spectroscopic studies of $Zn_{1-x}Mn_xO$ with a concentration ≤ 8 at.%, e.g. [Alaria 2005, Alaria 2006, Cong 2006, Gebicki 2005, Jouanne 2006, Phan 2007, Samanta 2007, Sato-Berru 2007, Wang 2005, Wang 2006/1, Xu 2006, Yang 2005]. Only few of them deal with Mn-implanted ZnO [Venkataraj 2007, Wang 2006/2, Zhong 2006]. Most of the Raman peaks reported for $Zn_{1-x}Mn_xO$ directly reflect the wurtzite lattice vibration modes of pure ZnO (subsection 3.1.2). In addition, features not present in the Raman spectra of pure ZnO occur. The origin of these features, such as isolated impurity modes, disorder-induced silent modes, or phonon modes by precipitates, is still controversial. In this subsection, systematic micro-Raman measurements from 10 K to room temperature are presented, using different excitation wavelengths throughout the visible spectral range, as reported in [Schumm 2008/1]. Besides Raman spectroscopy, X-ray diffraction (XRD) and transmission electron microscopy (TEM) are applied. Furthermore, the magnetic properties are probed by magnetooptical Kerr effect (MOKE) and electron paramagnetic resonance (EPR) experiments.

Sample	Fluence (cm^{-2})					Total Fluence (cm^{-2})
	450 keV	**300 keV**	**180 keV**	**100 keV**	**50 keV**	
0.2 at.%	$1.40 \cdot 10^{15}$	$5.25 \cdot 10^{14}$	$3.25 \cdot 10^{14}$	$2.13 \cdot 10^{14}$	$1.00 \cdot 10^{14}$	$2.56 \cdot 10^{15}$
0.8 at.%	$5.60 \cdot 10^{15}$	$2.10 \cdot 10^{15}$	$1.30 \cdot 10^{15}$	$8.50 \cdot 10^{14}$	$4.00 \cdot 10^{14}$	$1.03 \cdot 10^{16}$
2.0 at.%	$1.40 \cdot 10^{16}$	$5.25 \cdot 10^{15}$	$3.25 \cdot 10^{15}$	$2.13 \cdot 10^{15}$	$1.00 \cdot 10^{15}$	$2.56 \cdot 10^{16}$
8.0 at.%	$5.60 \cdot 10^{16}$	$2.10 \cdot 10^{16}$	$1.30 \cdot 10^{16}$	$8.50 \cdot 10^{15}$	$4.00 \cdot 10^{15}$	$1.03 \cdot 10^{17}$

Table 5.3: *Energies and fluences during the Mn^{2+} ion implantation.*

As mentioned above, commercially available ZnO single crystals (CrysTec, Berlin) were implanted at room temperature with manganese ions, yielding concentrations of 0.2, 0.8, 2, and 8 at.% relative to Zn. One reference sample was prepared by the implantation of Co with exactly the same ion energies and fluence as the 8 at.% Mn sample. In accordance with subsection 4.1.1

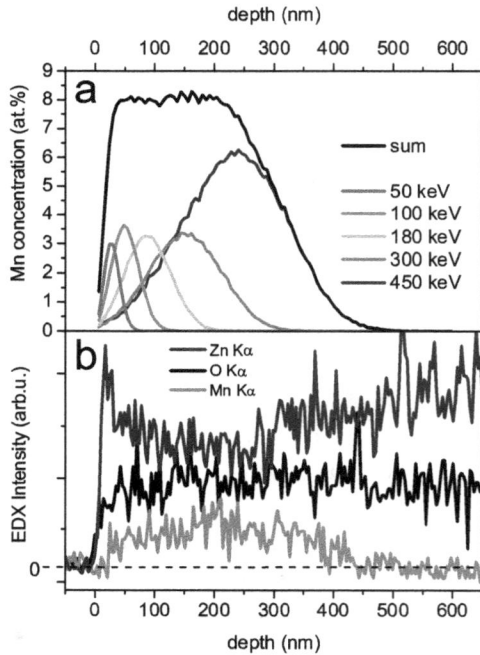

Figure 5.3: *Profile of manganese ions implanted in ZnO crystals with a concentration of 8 at.%, (a) calculated using SRIM/TRIM [Ziegler 1985] and (b) studied by EDX, from bottom to top: Zn K_α, O K_α, and Mn K_α (spectra have been scaled vertically for clarity).*

and section 5.1, a box-like implantation profile with a resulting layer thickness of about 300 nm was achieved by different ion energies (for energies and fluences see Table 5.3). The implantation profile displayed in Figure 5.3a was calculated with the Monte Carlo program package SRIM/TRIM [Ziegler 1985]. Post-implantation sample annealing was performed at 700 °C in air for 15 min. For the 8 at.% sample, a second annealing step, 900 °C in air for 15 min, was applied.

XRD analysis for this subsection was conducted with a Siemens D5000 diffractometer (Cu-K_α). Cross-section specimens for TEM analysis of the samples were prepared by a focused ion beam system (Novalab 600, FEI). Standard preparation procedures were used and finally the electron-transparent specimens were investigated in a high-resolution TEM Philips system (CM 200-FEG-UT) equipped with energy-dispersive X-ray spectroscopy (EDX).

EPR experiments were performed at the X band with 9.5 GHz on a Brucker ESP300 spectrometer using a microwave power of 2 mW. All data were taken at room temperature with magnetic field perpendicular to the crystal c-axis. The magnetic properties of the samples were also measured by the MOKE at room temperature with a linearly polarized He-Ne laser (632.8 nm) in the longitudinal configuration, probing the in-plane magnetization with a maximal field of 0.15 T. For the Raman measurements, both micro-Raman setups introduced in section 2.2 were used.

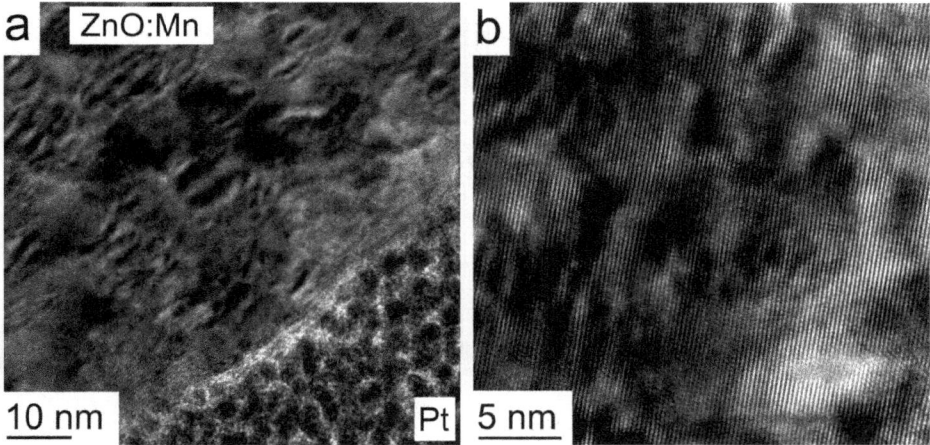

Figure 5.4: *(a) Cross-section TEM and (b) HRTEM images of 8 at.% Mn-implanted ZnO.*

Structural and magnetic characterization

XRD measurements show no new diffraction peaks for the Mn-implanted samples compared to the ZnO reference sample (not shown). Hence, within the sensitivity level of the XRD experiments, no clusters or other phases than ZnO have been formed upon the Mn implantation process or subsequent annealing. However, the (0002) diffraction of the ZnO crystal reveals a slight broadening, especially in the 8 at.% Mn sample, but no shift of this diffraction peak is observed. The high implantation flux leads to a high defect density, resulting in a locally distorted lattice and the broadened XRD peak.

The conducted TEM experiments confirm the assumption that no secondary phases have been created during ion implantation and subsequent annealing. Figure 5.4a and Figure 5.4b show the representative cross-section and the high-resolution TEM, respectively, of the sample implanted with the highest concentration of 8 at.%. The images demonstrate the presence of defects, seen by the strong variation of the contrast, but also the intact ZnO crystal lattice. Despite the investigation of large areas within the implanted region, no indication for additional Mn-Zn-O phases was found. EDX experiments confirm the calculated implantation profile (Figure 5.3). Slight differences in the implantation depth between the theoretical calculation and the experimental results may be due to limited depth calibration accuracy of the EDX setup. MOKE results reveal a weak paramagnetic signal at room temperature for the sample implanted with the highest ion fluence.

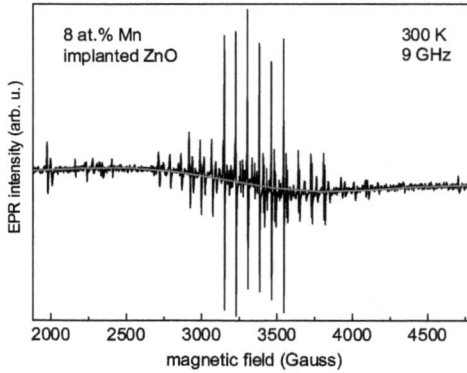

Figure 5.5: *EPR spectra of 8 at.% Mn-implanted ZnO: fine structure (black line) and broad background signal (red line).*

To get information about oxidation state and site occupancy of the Mn ions in the ZnO lattice, EPR measurements were conducted after the 700 °C annealing step. As the number of unpaired electrons is different for different oxidation states, they can be distinguished in the EPR spectra. Figure 5.5 shows the EPR signal for the 8 at.% Mn-implanted ZnO sample. A broad signal (red line in the figure) is superimposed by many narrow lines with a sextet having highest intensity. Similar EPR spectra have already been analyzed in detail [Schneider 1962/1, Schneider 1962/2]. The electronic configuration of the Mn^{2+} ion corresponding to its half-filled d-shell is $3d^5$ with spin $S = 5/2$. The only natural isotope is ^{55}Mn with nuclear spin $I = 5/2$. The resonance of an isolated Mn^{2+} ion located substitutionally on a Zn site in the hexagonal ZnO crystal is described by the following spin Hamiltonian:

$$\mathbf{H}_s = g\beta HS + \frac{1}{6}a\left(S_x^4 + S_y^4 + S_z^4\right) + D\left[S_z^2 - \frac{1}{3}S(S+1)\right] + ASI. \tag{5.1}$$

In Mn-alloyed ZnO with low Mn concentration (< 0.2 at.%), an isotropic Zeeman interaction (first term), a hyperfine interaction (last term), and a fine structure splitting (second and third term) are observed [Zhou 2003]. This signature, corresponding to Mn^{2+} ions located on Zn lattice sites, occurs already in the EPR spectra of the untreated ZnO crystals. Therefore, it can be stated that the ZnO crystals contain Mn^{2+} ions already as residual impurities, though with a concentration as low as about 10^{17} cm^{-3}. This Mn concentration is below the sensitivity limit of the Raman scattering method. For increased Mn concentrations (here, by implantation), an antiferromagnetic dipole-dipole interaction between substitutional Mn^{2+} ions occurs and the fine structure vanishes, which results in a broad, unstructured signal [Borse 1999, Zhou 2003, Zhou 2006]. Such a broad background signal is observed in the EPR spectra of the Mn-implanted ZnO for higher Mn concentrations (red line in the EPR spectra of 8 at.% Mn-implanted ZnO in Figure 5.5), caused by dipole-dipole interaction of the closely spaced Mn ions within the implanted layer. Thus, the total EPR spectrum consists of a fine-structured contribution by isolated, substitutional Mn^{2+} ions in the substrate and a broadened, unstructured signal from dipole-interacting, substitutional Mn^{2+} ions in the implanted layer.

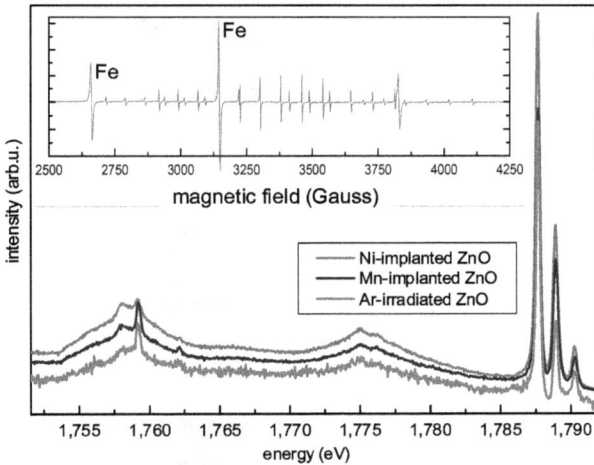

Figure 5.6: *Photoluminescence spectra of different ZnO host crystals due to transitions of residual Fe^{3+} impurities on Zn sites (excitation: $\lambda = 457.9$ nm, $T < 10$ K). Inset: EPR spectrum of 0.8 at.% Mn-implanted ZnO with additional features assigned to the residual Fe impurities as well.*

In the EPR spectra of the Mn-implanted ZnO crystals, additional signals were observed, which can be assigned to residual Fe impurities in the host crystals (inset in Figure 5.6). The intensity of these Fe impurity signals varied from sample to sample. Furthermore, during low temperature Raman experiments with the 457.9 nm line of the Ar ion laser as excitation, a structured luminescence was observed in the wavelength region between about 690 nm and 700 nm. As shown in Figure 5.6 for Ar irradiation as well as Mn and Ni implantation, this luminescence is not related to the implanted ions and was observed for all host crystals of this series. The luminescence between 1.75 eV and 1.80 eV can be identified as transitions of isolated Fe^{3+} ions on Zn^{2+} sites in the ZnO wurtzite lattice [Heitz 1992]. It was observed before in pure ZnO with residual Fe impurities as well as in Fe-implanted ZnO [Heitz 1992, Monteiro 2003]. Thus, besides the residual Mn impurities observed by EPR, also Fe impurities could be identified in the host crystals. In the series used for this thesis, however, the impurity concentration was very low, particularly, it was below the Raman spectroscopy sensitivity limit for such isolated ions. Nevertheless, for CrysTec crystals, which are widespread among the ZnO community, the residual impurities must be taken into account when conducting sensitive experiments.

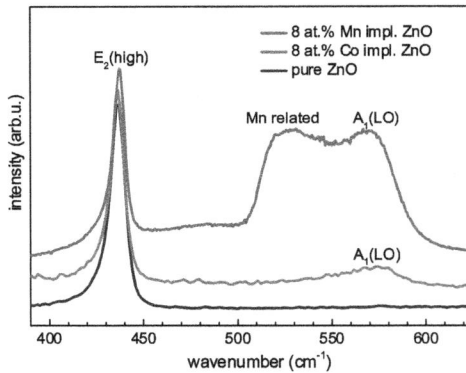

Figure 5.7: *Raman spectra of pure ZnO, 8 at.% Co, and 8 at.% Mn-implanted ZnO after 700 °C annealing (excitation: $\lambda = 514.5$ nm). The two-shoulder Raman signature between 500 cm^{-1} and 600 cm^{-1} is characteristic for Mn-alloyed ZnO.*

Mn-related additional modes: radiation damage, impurity-induced disorder, and isolated impurity modes

Figure 5.7 shows the Raman spectra of pure ZnO and ZnO implanted with 8 at.% Mn, compared to ZnO implanted with 8 at.% Co. In addition to the ZnO E_2(high) mode, the main new features in the spectra of the ZnO:TM samples are observed in the range of 500-600 cm^{-1}. In the case of the Co-implanted sample, a broad signal occurs at about 575 cm^{-1}. It can be assigned to the A_1(LO) mode of pure ZnO, intensified and broadened by TM-implantation-

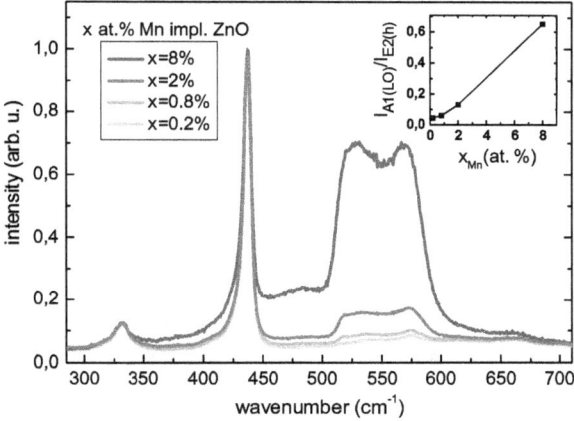

Figure 5.8: *Raman spectra of 0.2, 0.8, 2.0, and 8.0 at.% Mn-implanted ZnO after 700 °C annealing, normalized to the E_2(high) mode (excitation: $\lambda = 514.5$ nm). The intensity of the broad band between 500 cm^{-1} and 600 cm^{-1} scales with the Mn concentration. The inset shows the intensity ratio $I_{A1(LO)}/I_{E2(high)}$ versus the Mn concentration.*

induced disorder (see subsection 4.1.1). In the case of the Mn-implanted sample, an additional shoulder is visible in this region at lower wavenumbers, controversially discussed in the literature as partly or completely disorder related [see literature references concerning Raman scattering on ZnMnO above]. As shown in Figure 5.8, the intensity of the Raman signal in the range of 500-600 cm^{-1} is directly correlated with the Mn concentration. As will be shown below, two mechanisms participate in this effect. Firstly, the disorder generated by the radiation damage of the implantation process is of course higher for higher implantation concentrations. Secondly, the higher concentration of Mn ions in the ZnO crystal can induce impurity modes or disorder. Even if perfectly incorporated on Zn sites, the Mn ions induce disorder, inherent to the mixture of Zn and Mn on the cation sublattice. Besides the increased intensity of the phonon features and a pronounced peak broadening in Figure 5.8 for increasing concentration, there occurs also a rise of the background between 300 cm^{-1} and 700 cm^{-1}. It is especially obvious by comparison of the 4 at.% and the 8 at.% samples in the spectral region between 450 cm^{-1} and 500 cm^{-1}. Accordingly, the phonon density of states has a maximum in this region (subsection 3.1.2).

To evaluate the different contributions to the broad band in the Raman spectra, resonance measurements were conducted, using laser excitation wavelengths from 632.8 nm (1.96 eV) to

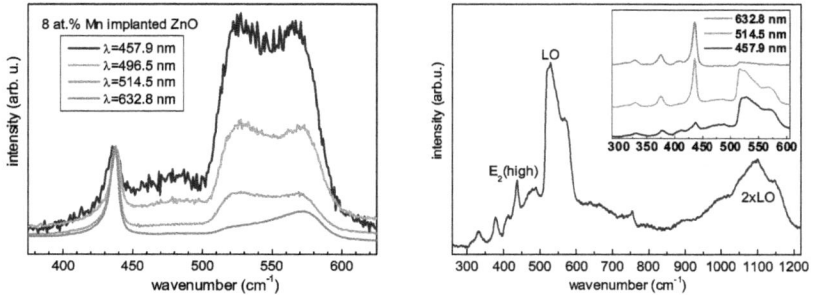

Figure 5.9: *(a) Resonance effect in the Raman spectra of 8 at.% Mn-implanted ZnO after 700 °C annealing. The spectra were normalized to the E_2(high) mode. Excitation from bottom to top: λ = 632.8, 514.5, 496.5, and 457.9 nm. (b) Resonance effect in the Raman spectra of polycrystalline bulk ZnMnO with 4 at.% Mn. Excitation: λ = 632.8 (red curve), 514.5 (green curve), and 457.9 nm (blue curves). The resonance predominantly affects the LO and the 2xLO regions.*

457.9 nm (2.71 eV). If the energy of the exciting laser approaches the energy of the ZnO band gap (3.4 eV or 365 nm), resonance effects occur in the Raman spectra of ZnO (see subsection 2.1.2 and section 4.1). In resonance Raman studies of pure, ordered ZnO and with excitation wavelengths between 457.9 nm and 647.1 nm, the strongest resonant enhancement was observed for the LO and the 2xLO region [Calleja 1977]. As discussed in subsection 2.3.2, the scattering cross section of LO phonon modes is affected by both deformation potential and Fröhlich scattering. The generally weak occurrence of the LO modes in Raman experiments far from the resonance can be attributed to the fact that these two mechanisms nearly neutralize each other [Calleja 1977]. When the excitation approaches the band gap energy, however, the strong LO resonance is mainly driven by the Fröhlich interaction of free excitons and zone center phonons, while the less pronounced TO resonance is due to interaction with continuum electron-hole states via a deformation potential [Scott 1969]. If the exciting light is exactly in resonance, the signal of the LO region and its overtones solely dominate the spectra [Bergman 2005, Scott 1970]. For decreasing laser wavelengths, a particularly strong rise is observed for the relative intensity of the broad band in the LO region (Figure 5.9a). This resonance effect can be explained by impurity-induced enhancement of the Fröhlich scattering. The Fröhlich interaction mechanisms under influence of impurities are described in [Kauschke 1987]. In Figure 5.9b, the according resonance effect of the LO and 2xLO region is shown for a polycrystalline, bulk ZnMnO sample. The impurity-induced resonance effect appears even stronger because no pure ZnO substrate can participate to the Raman spectra. In summary, the resonance results support the hypothesis that, besides the radiation damage effect, the Mn impurities play a major role in the formation of this broad band in ZnMnO.

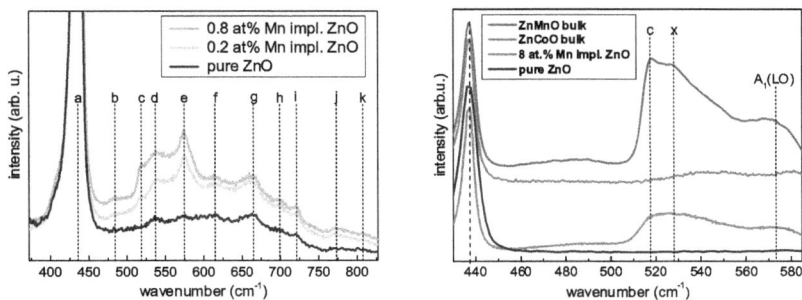

Figure 5.10: *(a) Raman spectra of pure ZnO, 0.2 at.%, and 0.8 at.% Mn-implanted ZnO after 700 °C annealing, normalized to the E_2(high) mode (excitation: $\lambda = 514.5$ nm for Mn-implanted ZnO and $\lambda = 632.8$ nm for pure ZnO). Due to the low Mn concentration, several features are resolved between 500 cm^{-1} and 750 cm^{-1}. (b) Raman spectra of pure ZnO, 8 at.% Mn-implanted ZnO after 700 °C annealing, bulk $Zn_{0.95}Co_{0.05}O$, and bulk $Zn_{0.96}Mn_{0.04}O$ (excitation: $\lambda = 514.5$ nm). The additional left shoulder is specific to Mn-alloyed ZnO. In the spectrum of the bulk ZnMnO sample, two features can be clearly observed within this shoulder.*

The above-mentioned two shoulder structure within the broad band between 500 and 600 cm^{-1} can be identified in all spectra of the annealed Mn-implanted ZnO samples. The right shoulder at about 575 cm^{-1} is seen in ZnO:TM for many different TM ions and can be assigned to the disorder-increased A_1(LO) mode. The results for V-, Fe-, Ni-, and Co-alloyed ZnO also confirm this behavior, see for example the Co-implanted reference sample in Figure 5.7 or sections 5.3 to 5.6. This effect is also seen in other ZnO samples with structural defects, e.g. in ZnO containing oxygen vacancies [Exarhos 1995] or in the disordered bulk samples and nanocrystals in chapter 4. Therefore, this right shoulder is not specific to Mn incorporation. While there are several Raman features identified for ZnO:TM in this spectral region, the strong and broad left shoulder at about 520-530 cm^{-1} is not observed for other TM in ZnO, see [Bundesmann 2003] and sections 5.3 to 5.6, for example. It is not known as a Raman signal of pure ZnO, either (subsection 3.1.2). Therefore, this feature can be identified as Mn related. For a more detailed mode discussion of this spectral region, Figure 5.10a provides the spectra of 0.2 and 0.8 at.% Mn-implanted ZnO after 700 °C annealing. The low Mn concentration and therefore lower disorder in those samples allow to resolve separate modes instead of the broad, unresolved band caused by the high phonon DOS in this region (subsection 3.1.2). The features at about 437 cm^{-1} (labeled a) and 575 cm^{-1} (labeled e) can be attributed to the modes E_2(high) and

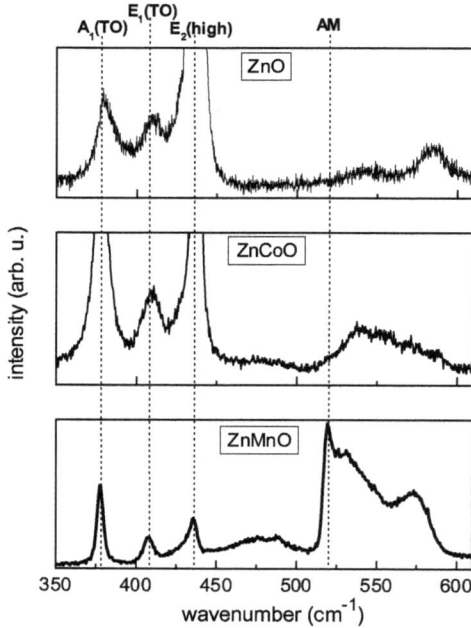

Figure 5.11: *Raman spectra of bulk, polycrystalline $Zn_{0.97}Co_{0.03}O$, $Zn_{0.96}Mn_{0.04}O$, and ZnO (excitation: $\lambda = 514.5$ nm). The additional mode (labeled AM) is clearly seen in this spectrum of Mn-alloyed ZnO in contrast to polycrystalline pure and Co-alloyed ZnO.*

$A_1(LO)$. By comparison to the modes presented in subsection 3.1.2, the mode at 483 cm^{-1} (labeled b) can be assigned to a zone-boundary multi-phonon process of ZnO. Furthermore, all features registered between 600 and 825 cm^{-1}, i.e. (f)-(k) in Figure 5.10a, are also seen in the measurements for the pure ZnO substrate. They can therefore be explained by Raman-active multi-phonon modes of pure ZnO. Some of these multi-phonon modes have a stronger intensity in the Mn-implanted samples than in pure ZnO due to disorder effects and impurity induced resonance. There are two more modes visible in 5.10a: at about 519 and 537 cm^{-1}, denoted by c and d, respectively. Both of these modes have been assigned to local vibration modes (isolated impurity modes) in the literature before [see references given above]. Nevertheless, the 537 cm^{-1} mode also occurs for pure ZnO. In addition, Cusco et al. ascribe this mode to a 2xB$_1$(low) or, alternatively, to a 2xLA process [Cusco 2007]. According to these results, only the mode at about 519 cm^{-1} in these spectra cannot be explained by intrinsic ZnO modes. Interestingly, this feature significantly increases in intensity with rising Mn concentration from the 0.2 at.% sample to the 0.8 at.% sample, which suggests a Mn impurity mode as origin.

In the following, the implanted layers are compared to the bulk $Zn_{0.96}Mn_{0.04}O$ sample mentioned above. This polycrystalline sample was grown by a vapor phase transport method at about 900 °C [Jouanne 2006]. In contrast to the samples implanted with Mn concentrations ≤ 0.8 at.%, not only the 519 cm^{-1} mode (labeled c in Figure 5.10b), but also a second additional feature can be identified in the left shoulder at about 527 cm^{-1} (labeled x). This additional feature is also indicated for the 8 at.% Mn-implanted ZnO sample. Note that also in other than the standard backscattering configuration, the additional shoulder is only present in Mn-alloyed ZnO, as can be seen by a comparison to polycrystalline bulk ZnO and ZnCoO samples in Figure 5.11. The different growth process and the high growth temperature of the bulk sample may lead to a better substitutional incorporation of the Mn ions into the ZnO lattice and therefore to the pronounced Mn-related features.

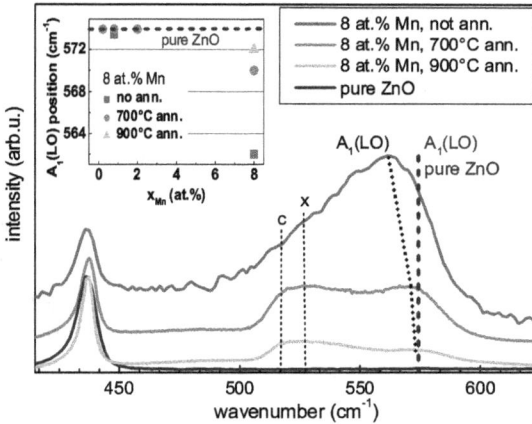

Figure 5.12: *Raman spectra of pure ZnO and 8 at.% Mn-implanted ZnO, unannealed, after 700 °C, and after 900 °C annealing (excitation: $\lambda = 514.5$ nm). The inset shows the $A_1(LO)$ mode position in dependence on the Mn concentration and the applied annealing steps.*

In order to study such temperature effects and to get information about the origin of the 527 cm^{-1} and the 519 cm^{-1} modes, an annealing sequence was performed for the implanted layers with temperatures up to 900 °C. Such annealing procedures are typically applied for implanted samples in order to heal radiation damage (see section 5.1) and to support a substitutional incorporation of the implanted ions. Pure, but disordered ZnO exhibits a strong, disorder-enhanced $A_1(LO)$ mode and a broadened E_2(high) peak. Annealing of such a sample could improve the crystalline quality and, consequently, the $A_1(LO)$ mode would be reduced and the half-width of the E_2(high) phonon mode would become narrower. The commercial, pure ZnO substrate samples, used for the implantation, are of excellent crystal quality and exhibit a very strong, narrow E_2(high) mode and only a very small $A_1(LO)$ signal before implantation. Their crystal quality is not significantly improved by annealing, so the Raman spectra of these pure ZnO samples look identical without annealing, after 700 °C, and after 900 °C annealing. Figure 5.12 shows the 8 at.% Mn-implanted sample without annealing, after annealing for 15 min at 700 °C, and after an additional annealing for 15 min at 900 °C, respectively. For comparison, also the spectrum of pure ZnO is shown. The unannealed sample shows an extremely strong, broad band between 500 and 600 cm^{-1} with a dominating right shoulder. Its maximum is at about 560 cm^{-1} and thus about 14 cm^{-1} away from the 574 cm^{-1} position which can be expected for the $A_1(LO)$ mode in well-ordered ZnO. This strong intensity of the $A_1(LO)$ mode and the shift to lower wavenumbers can be interpreted as a consequence of the radiation damage in this unannealed sample. The shift is in accordance with the dispersion of the $A_1(LO)$ mode near the Brillouin zone center (subsection 3.1.2).

Upon annealing at 700 °C, the characteristic two-shoulder structure can be seen, mostly due to the strong reduction of the broad $A_1(LO)$ mode. This is confirmed by the second annealing step at 900 °C, which results in another strong reduction of the $A_1(LO)$ structure. After this 900 °C annealing step, the left shoulder shows a slight shift to lower wavenumber values while approaching two-feature behavior (at about 519 cm^{-1} and about 527 cm^{-1}) as identified for the bulk $Zn_{0.96}Mn_{0.04}O$ sample in Figure 5.10b. Both modes, at about 519 cm^{-1} and 527 cm^{-1}, are reported in the literature for Mn-alloyed ZnO, but without effort to distinguish between these modes. Nevertheless, both are equally taken as evidence for a substitutional incorporation of Mn on Zn sites [Alaria 2006, Du 2006, Venkataraj 2007, Xu 2006]. However, it will be shown now that each of these modes has its own origin. The feature at 527 cm^{-1} is strong in the Raman spectra of the more disordered samples with Mn concentrations ≥ 2 at.% (Figures 5.7 and 5.8) and decreases upon annealing (Figure 5.12). Besides, a mode at 528 cm^{-1} is also observed in Sb-alloyed ZnO [Bundesmann 2003, Zuo 2001], an evidence that it is no isolated impurity mode, but probably a ZnO mode, activated or intensified by Mn ion incorporation. For example, Manjon et al. calculated the $2xB_1$(low) mode to be located in this spectral region [Manjon 2005]. In contrast, the feature at 519 cm^{-1} is covered by the broad disorder band in the spectra of highly disordered Mn-alloyed ZnO and therefore it is only seen for the samples with Mn concentrations ≤ 0.8 at.% or after thermal treatment up to 900 °C. Its intensity scales with the Mn content in such well-ordered samples, as shown in Figure 5.10a for the samples with 0.2 and 0.8 at.% Mn. Additionally, the EPR experiments confirm that the Mn ions are mainly substitutionally incorporated in the annealed samples, and therefore encourage the assumption that this mode is a candidate for an isolated impurity mode of substitutionally incorporated Mn in ZnO.

Still, there are only very few theoretical studies of isolated impurity vibration modes by TM ions in ZnO [Bundesmann 2005, Lakshmi 2006, Thurian 1995, Zhong 2006]. Moreover, they do not deliver a clear and consistent picture regarding their existence and the expected frequency values. In section 2.3, additional modes by isolated impurities in a crystal lattice were classified in local vibration modes, gap modes, and band modes. For a local vibration mode, an impurity vibration with wavenumber values above 600 cm^{-1}, i.e. above the optical phonon branches, would be required from the isolated Mn in ZnO. Such high frequencies can not be expected for Mn which has a similar mass as the substituted Zn ion ($m_{Zn} \approx 65$ u, $m_{Mn} \approx 55$ u). Besides, the identified feature is located well below that value. A gap mode would lie between 260 cm^{-1} and 370 cm^{-1}, i.e. between the acoustic and the optical phonon branches. Hence, in this simplified picture, only a band mode could explain the observed feature. A band mode, however, corresponds to a vibration induced by the isolated impurities, but not localized at their sites. On the other hand, as mentioned in section 2.3, the situation is more complicated in anharmonic crystals. Additionally, there is also the possibility that the vibration does not occur due to isolated Mn impurities, but due to Mn-related complexes in ZnO. Though, no indications for such complexes were found in the examined samples with low concentrations.

In summary, the overall intensity of the left shoulder at about 515-530 cm^{-1} in the Raman spectra of ZnO:Mn should not be taken uncritically as evidence for a substitutional incorporation of Mn on Zn sites or for an estimation of the actual Mn content. Instead, the two different modes identified within this shoulder have to be distinguished carefully. In future research, an enhanced theoretical understanding and also additional experimental methods, such as X-ray photo-electron spectroscopy (XPS) or X-ray absorption spectroscopy (XAS), could help to further clarify the exact origin of the identified additional Raman features.

Depth profile analysis of the Mn-implanted layers

To study the depth profile of the Mn distribution and the associated disorder effects within the implanted layer, a micro-Raman depth analysis was conducted. In this depth scan, the laser spot position of the Raman microscope was stepwise varied within the implanted layer and the underlying part of the substrate (total thickness about 0.5 mm). The absolute intensity of the various features in the Raman spectrum at the different steps of the depth scan is determined by the convolution of the sample's response function and the optical detection profile. The detection profile depth amounts to 6 ± 0.5 μm FWHM. In Figure 5.13, the results of the depth scans are shown for the 8 at.% sample unannealed and after the second annealing step (900 °C), respectively. In all spectra, the vertical scale is normalized to the ZnO E$_2$(high) mode intensity. This implies a strong stretching of up to 40 times for focus positions above the sample surface

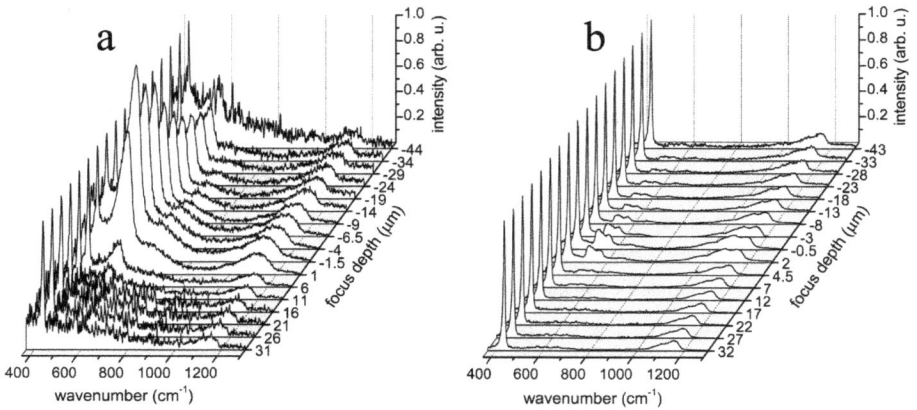

Figure 5.13: *Depth-dependent Raman spectra of the 8 at.% Mn-implanted ZnO sample, (a) unannealed and (b) 900 °C annealed. The spectra are normalized to the E_2(high) mode (excitation: $\lambda = 514.5$ nm). '0' at the focus depth axis corresponds to focusing on the sample surface, negative values denote focus positions above the surface (air), positive values below (within the sample).*

(negative depth values in Figure 5.13). Stray light caused by diffuse reflection of the laser beam at the sample surface dominates the spectrum of the unannealed sample for high negative focus depths. Comparable stray light noise is not observed for the annealed samples. This confirms that the implantation damage on the surface and in the near surface region is strongly reduced by the applied annealing. Deconvolution of the Raman depth scans indicates that the signals identified as related to the implantation derive from a near-surface region.

The spectra of the unannealed sample are dominated by the disorder-intensified A_1(LO) mode. In contrast, the spectra of the 900 °C annealed sample show the two shoulder signature discussed above. Note that in the latter spectral series the Mn-related left shoulder is particularly strong, compared to the disorder-related right shoulder, exactly at the focus depth for which the shoulder reaches its overall intensity maximum. This is a further indication that the two signals have different origins and obviously also a different depth distribution. In Figure 5.13, also a broad band between 900 and 1300 cm^{-1} is visible which originates from second-order processes. This band was not discussed before in detail for Mn-alloyed ZnO. It is dominated by features at about 1160, 1100, and 990 cm^{-1}, which are also reported for pure ZnO (subsection 3.1.2). Additionally, a broad feature evolves at about 1060 cm^{-1} which does not occur in pure ZnO. Its intensity is correlated with the intensity of the feature at 527 cm^{-1}. Therefore, it is assigned to the second-order process of the 527 cm^{-1} mode.

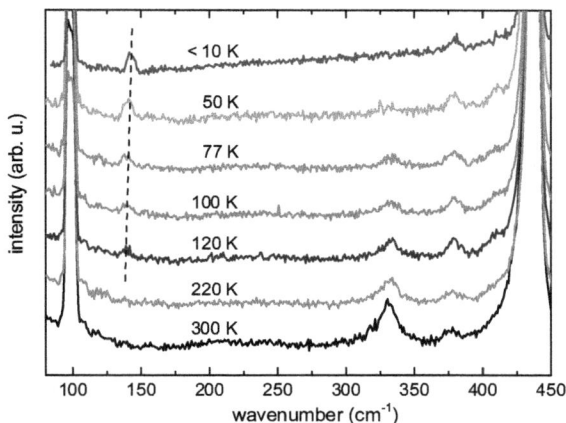

Figure 5.14: *Temperature-dependent Raman spectra of the 0.2 at.% Mn-implanted ZnO sample after 700 °C annealing (excitation: $\lambda = 457.9$ nm). While the difference mode at about 330 cm^{-1} disappears at low temperature, an additional mode appears for $T \leq 120$ K at about 140 cm^{-1}. Note the weak intensity of this additional mode compared to the E_2(high) and E_2(low) modes at about 437 cm^{-1} and 100 cm^{-1}, respectively.*

Secondary phase discussion and low temperature additional mode

While Mn incorporation is the key to the desired magnetic properties of the DMS $Zn_{1-x}Mn_xO$, one must also take into account the possibility of precipitate formation. Ferromagnetic Mn_3O_4, antiferromagnetic elemental Mn, or antiferromagnetic MnO could be reasons for the often contradictory reports about the magnetic properties of Mn-substituted ZnO. The predominating Raman modes for MnO, MnO_2, Mn_2O_3, $ZnMn_2O_4$, and Mn_3O_4 are in the spectral range between 300 cm^{-1} and 700 cm^{-1} [Buciuman 1999, Garcia 2005, Julien 2004]. An enhanced sensitivity for secondary phases as well as for localized impurity modes can be expected in low temperature Raman measurements due to more pronounced and sharper phonon peaks. In the low temperature spectra of Figure 5.14, two features exhibit a temperature dependence. First, the difference mode E_2(high)-E_2(low) at 330 cm^{-1} disappears for low temperatures (section 4.1). Second, a weak additional mode appears at about 139 cm^{-1}. Towards Helium temperature, the latter shifts to higher wavenumber values until about 143 cm^{-1} and becomes sharper and more intense. This feature does not correspond to any characteristic mode of the Mn-related precipitates mentioned above. In addition, a similar mode was identified in low temperature Raman spectra of pure

ZnO [Cusco 2007]. Because of its temperature behavior it may be related to the vibration of a localized, intrinsic ZnO defect. So again, no secondary phases could be identified in the analyzed ZnMnO samples with concentrations ≤ 8 at.%. This in accordance with the results by TEM and XRD discussed above and with literature findings regarding the solubility limit of Mn in ZnO [Jin 2001, Kolesnik 2004]. The antiferromagnetic coupling of the Mn on Zn sites, detected by EPR, can therefore be attributed to Mn ions substitutionally incorporated on Zn sites in ZnO and not to the formation of antiferromagnetic precipitates such as elemental Mn or MnO.

5.2.2 $Zn_{1-x}Mn_xO$ layers with concentrations ≥ 16 at.%

The fabrication of the 16 at.%, 24 at.%, and 32 at.% Mn-implanted samples studied in this subsection and the impact of the implantation on their crystal quality are discussed in section 5.1. While no secondary phases were observed in Mn-implanted ZnO with concentrations ≤ 8 at.% (subsection 5.2.1 and [Schumm 2008/1]), they are clearly present for the samples implanted with 16 at.%, 24 at.%, and 32 at.% after annealing at 700 °C. The additional phonon modes occurring in the Raman spectra of these samples (Figure 5.15 and Figure 5.16) can be attributed to the formation of Zn-Mn oxides [Buciuman 1999, Julien 2004, Samanta 2007]. In detailed lateral mapping of the sample implanted with 16 at.% Mn, oxide phases were only detected for few inclusions of about µm size on the sample surface (coverage < 0.1%). These inclusions appear orange under the microscope and show strong additional Raman signals. A similar surface structure was observed for the sample implanted with 24 at.% Mn. However, in this sample some of the inclusions are located in areas extended over several µm. Figure 5.15 shows such a secondary phase area appearing as a circle-like structure. Inside the circle, mostly ZnO is observed, with a crystal quality comparable to the large-area ZnO outside. In contrast, in the rim region with most of the secondary phase inclusions, ZnO with clearly enhanced crystal quality is detected. This indicates a self-purification process of the ZnO by forming oxide phases. The sample implanted with 32 at.% Mn shows a substantially different surface structure (Figure 5.16). Small precipitate islands of µm size are littered over the whole surface (coverage $\sim 30\%$). They appear green under the microscope, while the surrounding ZnO surface is red. The Raman spectra of the local precipitates in the 16 at.% and the 24 at.% sample as well as the Raman spectra of these green islands show three additional peaks at about 320 cm^{-1}, 380 cm^{-1}, and 680 cm^{-1}. From these frequency positions and their intensity ratios, they can be attributed to $ZnMn_2O_4$, which possesses three characteristic phonon modes at 327 cm^{-1}, 389 cm^{-1}, and 680 cm^{-1} [Samanta 2007]. The shifting of the additional Raman features compared to the literature values may be due to strain effects. In this case, the red shifts of up to 10 cm^{-1} indicates tensile strain of the oxide precipitates. Another plausible explanation for the peak shifts is the presence of non-stoichiometric $Zn_xMn_{3-x}O_4$ phases, as discussed below.

Because of their small size and very low coverage, the precipitates in the 16 at.% and the 24 at.% samples are below the XRD detection limit. The $ZnMn_2O_4$ islands of the 32 at.% Mn-implanted ZnO, however, are reflected in the corresponding diffractogram in Figure 5.17. The strong diffraction peaks at about 34.4° and about 72.6° are the 0002 and 0004 Bragg peaks of the hexagonal ZnO host crystal [JCPDS 1997]. The left shoulder at about 32.4° in the 0002 ZnO peak is observed for all samples and does therefore not depend on the TM species. Four additional peaks are observed in the diffractogram of this sample, which are attributed to

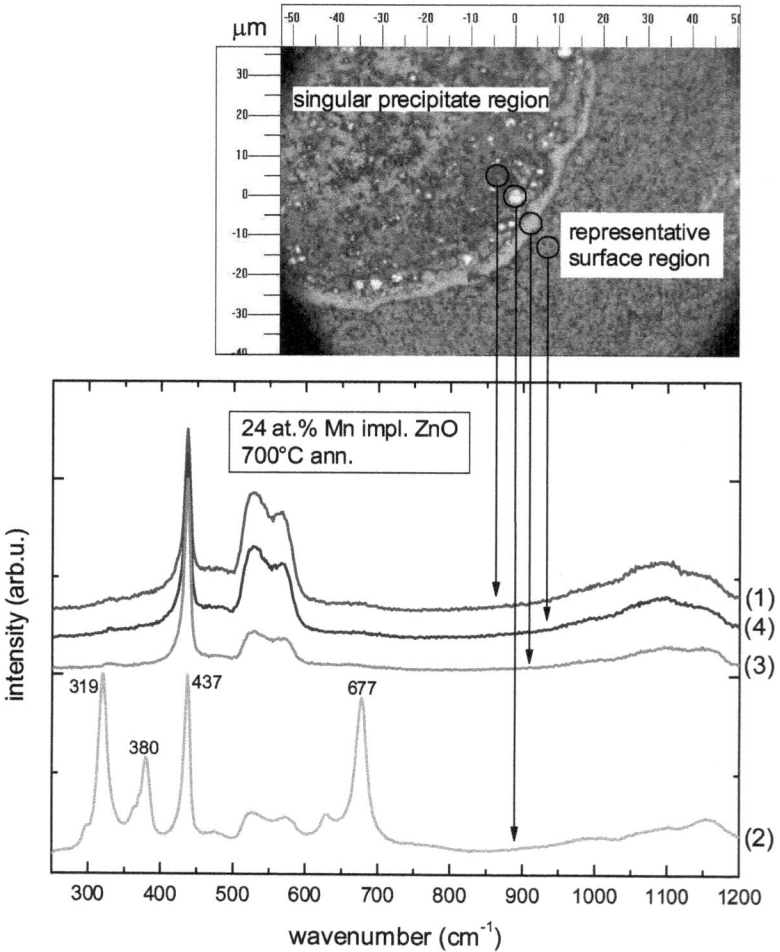

Figure 5.15: *Raman spectra of different spots on the 24 at.% Mn-implanted ZnO sample after 700 °C annealing show the inhomogeneity of the sample caused by precipitate formation (excitation: $\lambda = 514.5$ nm). The optical microscope picture shows the studied surface spots. Spectra: laser focused on (1) dark spot, (2) yellow spot, and (3) grey rim spot within the singular precipitate region; (4) spot on the representative surface region.*

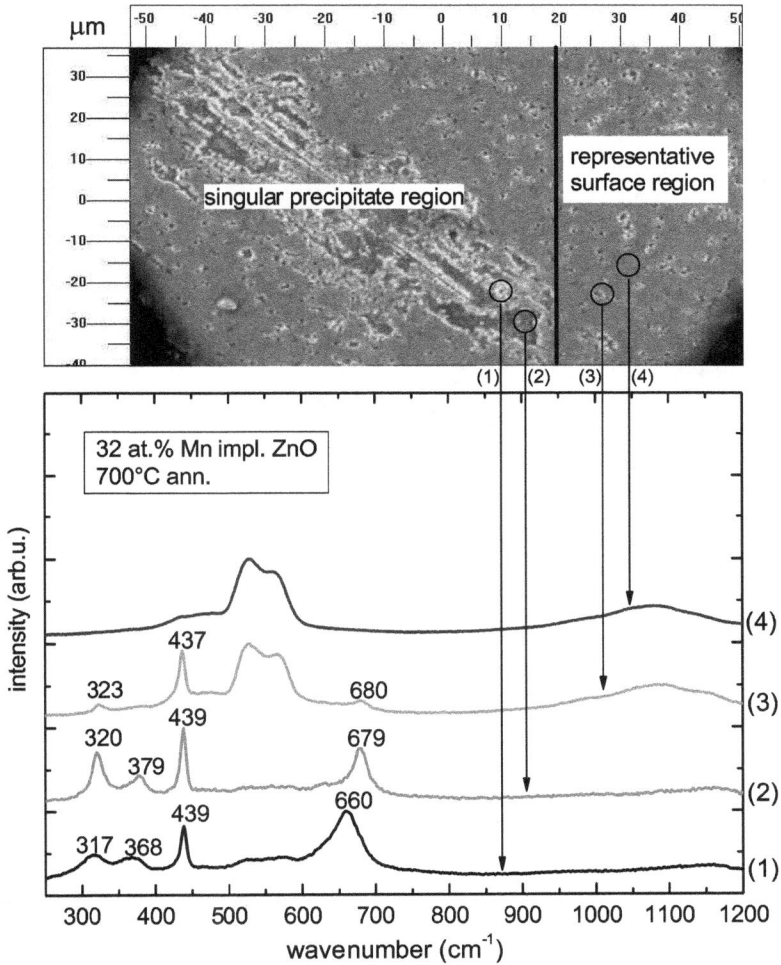

Figure 5.16: *Raman spectra of different spots on the 32 at.% Mn-implanted ZnO sample after 700 °C annealing show the inhomogeneity of the sample caused by precipitate formation (excitation: λ = 514.5 nm). The optical microscope picture shows the studied surface spots. Spectra from bottom to top: laser focused on (1) violet spot and (2) dark spot in the shown singular precipitate area; (3) green island and (4) red surface spot in the representative surface region.*

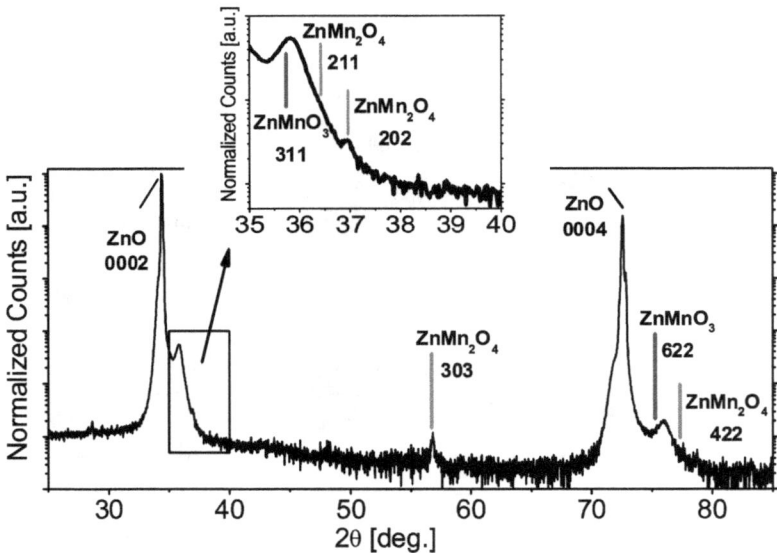

Figure 5.17: *XRD diffractogram of the 32 at.% Mn-implanted ZnO sample. Besides the ZnO Bragg peaks, four additional features are observed. They are assigned to the 202 and 303 peaks of $ZnMn_2O_4$ and to either the 211 and 422 peaks of $ZnMn_2O_4$ or the 311 and 622 peaks of $ZnMnO_3$ [JCPDS 1997].*

$ZnMn_xO_y$ formation, taking into account the Raman results. The diffraction peaks at about $37.0°$ and $56.8°$ correspond very well to the literature values of the 202 ($36.93°$) and the 303 ($56.74°$) Bragg peaks of $ZnMn_2O_4$ [JCPDS 1997]. With the relation of Scherrer, the size of the $ZnMn_2O_4$ segregations could be estimated to about 20 nm [Scherrer 1918].

The two stronger additional signals at about $35.8°$ and $75.9°$ are shifted to lower diffraction angles, but still close to the reported values of the $ZnMn_2O_4$ 211 ($36.40°$) and 422 ($77.34°$) reflections [JCPDS 1997]. Lattice mismatch between the host matrix and the secondary phase grains can be the source of local strain, which induces a variation of the lattice constant of secondary phase segregations [Cullity 1978]. The observed shift to smaller diffraction angles can therefore be explained by tensile strain effects, which is in accordance with the observed frequency shifts of the phonons modes discussed above. Another plausible explanation is the presence of $ZnMnO_3$, which shows its strong 311 and 622 reflections at $35.67°$ and $75.55°$, respectively [JCPDS 1997]. No literature data is available for Raman scattering on $ZnMnO_3$ to confirm this assignment. There is a XRD report of non-stoichiometric $Mn_{3-x}Zn_xO_4$ phases, which show a

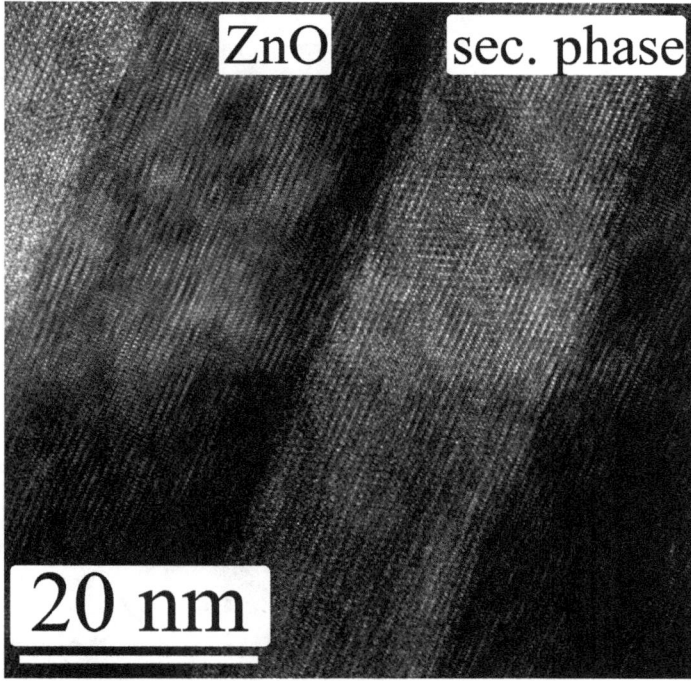

Figure 5.18: *High-resolution TEM picture of secondary phase clusters on the surface of the 32 at.% Mn-implanted sample after 900 °C annealing in air.*

strong reflection close to the observed 35.8° [Blasco 2006]. While non-stoichiometric precipitates are also suggested by the Raman data, no reflection corresponding to the 75.9° peak is reported for such phases.

Additionally, few areas were detected by micro-Raman mapping on the sample surface of the 32 at.% Mn-implanted sample, which contain islands appearing violet under the microscope (picture in Figure 5.16). Because of the local appearance of these inclusions, they are not expected to influence the XRD results. The additional Raman modes of these inclusions lie at 317 cm^{-1}, 368 cm^{-1}, and 660 cm^{-1}, which is in between the reported Raman features of stoichiometric ZnMn$_2$O$_4$ at 327 cm^{-1}, 389 cm^{-1}, and 680 cm^{-1} and of stoichiometric Mn$_3$O$_4$ at 310 cm^{-1}, 357 cm^{-1}, and 653 cm^{-1} [Julien 2004, Samanta 2007]. The intensity ratios of the peaks and the peak positions lie closer to Mn$_3$O$_4$ than to ZnMn$_2$O$_4$. These findings could be explained by Mn$_3$O$_4$ under compressive strain. But also a mixed mode behavior of non-stoichiometric Zn$_x$Mn$_{3-x}$O$_4$ with mode positions between the stoichiometric oxides Mn$_3$O$_4$ and ZnMn$_2$O$_4$ could account for these results, as it is known that non-stoichiometric Zn$_x$TM$_{3-x}$O$_4$ crystals are very often grown when preparing spinels with a nominal x = 1 composition [Piekarczyk 1988].

In addition to Raman scattering and XRD, the cluster formation on the surface of the 32 at.% Mn-implanted sample was studied by HRTEM and EDX line scan experiments after 900 °C annealing in air. The experimental details are described in subsection 5.2.1. In Figure 5.18, an example for the recorded TEM pictures is shown. The strong variation of the contrast can again be attributed to structural disorder of the implanted ZnO. Moreover, secondary phase clusters with different crystal structure and sub-μm size are clearly visible on the surface of the sample. Large areas of the sample surface were scanned and most of the observed precipitates were elongated and aligned to the wurtzite structure of the subjacent ZnO as shown in Figure 5.18. EDX line scans in the precipitate regions identified by HRTEM confirm a substantially higher TM concentration and decreased Zn concentration within the clusters.

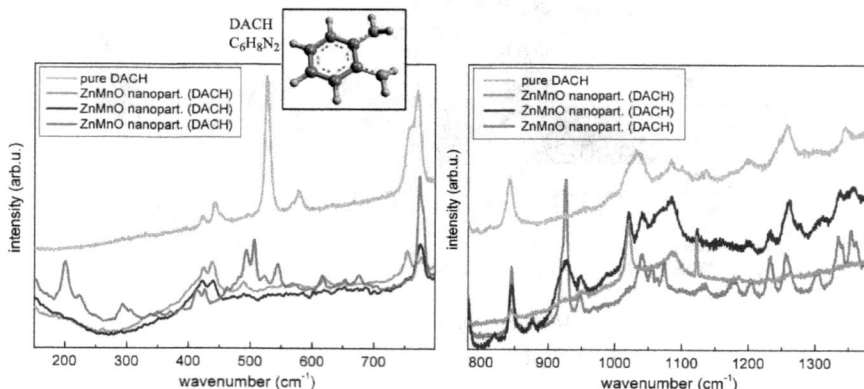

Figure 5.19: *Raman spectra of pure DACH and of three different ZnMnO nanoparticle samples fabricated by the same synthesis with DACH as capping ligand (excitation: λ = 514.5 nm).*

5.2.3   ZnO:Mn nanoparticles

Among the Mn-alloyed ZnO systems with reported ferromagnetism at room temperature, there are also ZnMnO nanoparticles [Kittilstved 2005, Wang 2006/1]. To study such systems, Mn-alloyed ZnO nanoparticles were fabricated based on the syntheses described in [Chory 2007] and discussed in section 4.2. Mn was added in the form of the precursor $MnCl_2$ with about 2 at.% Mn^{2+} relative to Zn^{2+} in order to obtain $Zn_{0.98}Mn_{0.02}O$ nanocrystals stabilized by organic ligands. The resulting particles were characterized by Raman scattering, EPR, and SQUID measurements.

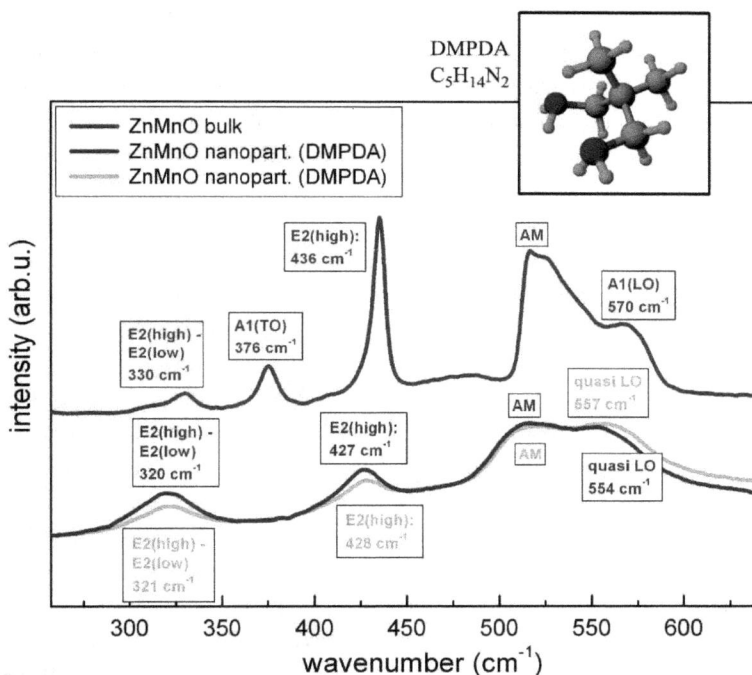

Figure 5.20: *Raman spectra of bulk ZnMnO and of two different ZnMnO nanoparticle samples fabricated by the same synthesis with DMPDA as capping ligand (excitation: $\lambda = 514.5$ nm).*

For the first synthesis procedure, the organic molecule DACH (diaminocyclohexane) was chosen as capping ligand (see inset in Figure 5.19) because good results were achieved for pure ZnO nanoparticles using this stabilizer. Figure 5.19 shows the Raman spectra of the nanoparticles resulting from this synthesis and also the spectrum of pure DACH. The three nanoparticle curves correspond to three different sample batches, fabricated by the same synthesis procedure to check reproducibility. However, no signal corresponding to wurtzite ZnO is observed for the DACH-stabilized particles. Furthermore, the feature-rich spectra are obviously not solely due to vibrations of the DACH ligand. By the observed vibration frequencies, also residues of the TMAH agent can be ruled out, which were identified for several syntheses of pure nanoparticles in section 4.2. A possible explanation for the Raman results are vibrations of clusters including precursor material, Mn ions, and the DACH molecules.

As a more promising ligand candidate, DMPDA (dimethylpropylenediamine) was identified (see inset in Figure 5.20). Using DMPDA as capping ligand, a strong, chelate-type bonding involving both amino groups of DMPDA can be expected for the nanoparticle-molecule interface. Figure 5.20 shows the Raman spectra of the DMPDA-stabilized, Mn-alloyed nanoparticles and of polycrystalline, bulk $Zn_{0.96}Mn_{0.04}O$, for comparison. The bulk sample was described in subsection 5.2.1. Again, the two nanoparticle curves correspond to two different batches synthesized by the same procedure. The spectra of the nanoparticles show the ZnO phonon modes $E_2(high)-E_2(low)$ and $E_2(high)$, but also the two-shoulder structure between 500 cm^{-1} and 600 cm^{-1}, which is characteristic for Mn-alloyed ZnO. Additionally, all Raman features of the nanoparticles show a pronounced red shift and strongly broadened peaks. Different contributions to this effect are plausible: Relaxation of the selection rules due to increased crystal disorder, local heating, and alloy potential fluctuations due to the Mn impurities. Crystal disorder and local heating as cause of peak shifting and broadening were already discussed in detail in sections 2.3 and 4.2. As the DMPDA-capped particles were studied with a high laser power density (>1000 kW/cm^2, Dilor system), local heating is assumed to have a strong impact especially on the peak position. For such high power densities, shifts of more than 10 cm^{-1} were observed in ZnO nanoparticles [Alim 2005/1, Alim 2005/2]. In contrast, topological crystal disorder, like interstitials, stacking faults, etc., is not likely to cause such pronounced red shifts. In alloyed semiconductors, however, also the so-called alloy potential fluctuations (APF) due to atom substitutions result in a reduced symmetry [Parayanthal 1984]. Wang et al. used a simple spatial correlation model to describe experimentally observed peak shifts and broadening in the Raman spectra of ZnMnO nanoparticles with average crystal diameter of ~50 nm [Wang 2005]. Thereby, they observed and calculated peak shifting of 2 cm^{-1} and broadening of 11 cm^{-1} for the $E_2(high)$ mode in nanoparticles with 2 at.% Mn.

The $E_2(high)$ mode in the Raman spectra of the DMPDA-stabilized ZnO nanoparticles shows a red shift of about 9 cm^{-1} and a broadening of about 15 cm^{-1} compared to pure ZnO nanoparticles fabricated by a similar synthesis procedure (Figure 5.20). Similar values are observed for the quasi-LO and the $E_2(high)-E_2(low)$ modes. To further study the impact of crystal disorder on the shifting and broadening, annealing experiments were performed on the DMPDA-stabilized nanoparticles for 30 min at 350 °C in air. Furthermore, the Raman spectra were taken at very much lower laser power density (<30 kW/cm^2, Renishaw system) to study the Raman scattering without strong local heating. Figure 5.21 shows the results of these experiments. Obviously, already the spectra of the unannealed particles are different from the above presented spectra. They show comparable red shifts, but the peak broadening is much weaker than in the spectra recorded with higher laser power. Due to the thermal treatment, the Raman peaks become even narrower and exhibit red shifts of only <5 cm^{-1} compared to the literature values for bulk ZnO. Additionally, the broad unstructured disorder band between 500 cm^{-1} and 600 cm^{-1} evolves into the well-known two-shoulder signature, which is characteristic for Mn-alloyed ZnO. This

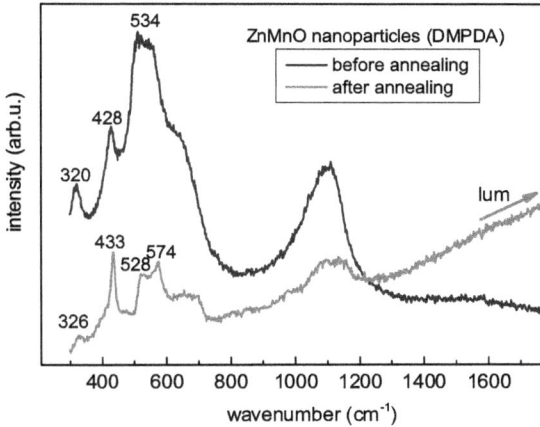

Figure 5.21: *Raman spectra taken before and after annealing experiments on the ZnMnO nanoparticles capped with DMPDA (excitation: $\lambda = 514.5$ nm). During the thermal treatment, the particles were placed on a heat plate at $350\,^\circ C$ for 30 min.*

structure had been observed in Figure 5.20 without thermal annealing, but at such high laser power densities that a laser-induced healing effect seems plausible.

The results indicate that the observed peak broadening in the Raman spectra of the DMPDA-capped ZnMnO nanoparticles is mostly temperature-driven, while the red shift is caused by crystal disorder, impurity-induced APF, and local heating. However, the thermal treatment not only improves the crystal structure of the particles, but can also harm the organic ligands. This could be the cause for a luminescence signal occurring in the Raman spectra after the annealing as indicated in Figure 5.21. Furthermore, damaging the capping molecules could also lead to a clustering of the ZnO crystallites. In this case, the heat conduction could be increased, further decreasing the impact of local heating.

In such wet-chemically synthesized particles, the position of the TM ion is even more ambiguous than in TM-implanted ZnO or ZnO:TM grown by VPT techniques. Besides the substitutional or interstitial position within the ZnO lattice, the TM ions could also bind to the organic ligands outside the nanocrystals. In the Raman spectra of Figure 5.20, however, no indications for Mn-organic clusters are found as they were observed in the case of the DACH-stabilized particles. Furthermore, also no secondary phases are detected. To get further information on the Mn ion position in the DMPDA-stabilized nanoparticles, they were studied by EPR (for experimental

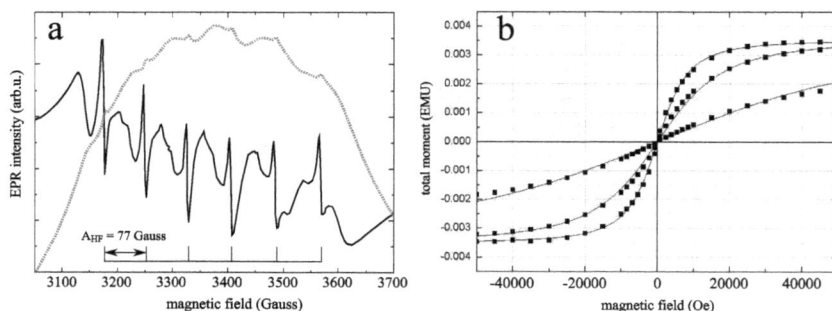

Figure 5.22: *(a) EPR spectra of DMPDA-capped ZnMnO nanoparticles: derivative spectrum (black line) and integrated spectrum (red line). (b) Corresponding SQUID measurements of the magnetization as a function of the magnetic field taken at 2, 4.3, and 15 K (squares). The data points are fitted with a magnetic moment of J = 4 (lines).*

details, see subsection 5.2.1). In Figure 5.22a, again the six lines attributed to isolated, substitutional Mn^{2+} are observed in the derivative mode, as discussed for bulk ZnMnO in subsection 5.2.1. However, the spectra of nanocrystals are more complicated. Only the lattice sites in the nanocrystal core can be described with bulk symmetry. The symmetry is reduced for lattice positions closer to the nanocrystal surface, resulting in relaxed selection rules. This causes additional lines in the EPR spectra, which are located in between the six-line pattern. It should be noted that for Mn^{2+} ions located on the surface of nanocrystals, a larger hyperfine interaction has been observed [Kennedy 1995, Zhou 2003]. This is caused by the reduced covalent bonding of the surface-Mn^{2+} compared to the Mn ions located in the core of the nanocrystals. However, this contribution is not observed in the spectra of the nanoparticles with DMPDA as capping ligand.

While the six-line signature in the derivative spectrum (black line in Figure 5.22a) proves the existence of isolated, substitutional Mn ions, the integrated spectrum (red line) additionally indicates the interaction of nearby Mn ions within the particles or in clusters.

The different Mn ion positions identified by EPR are also reflected in the magnetic properties of the nanoparticles. While isolated Mn ions have magnetic moments of $J = 5/2$, the paramagnetic behavior of the particles determined by SQUID measurements can be fitted with $J = 4$, see Figure 5.22b. Note that elemental Mn is antiferromagnetic, Mn_3O_4 is ferromagnetic, and most other Mn oxides are antiferromagnetic.

In summary, the DMPDA-capped nanoparticles exhibit the Raman fingerprint of ZnMnO, whereas peak shifts and broadening can be explained by local heating effects, crystal disorder, and alloy potential fluctuations. No secondary phases or clusters are detected by Raman scattering. Mn ions are positioned substitutionally within the crystal, but also magnetic interaction of nearby Mn ions is observed by EPR.

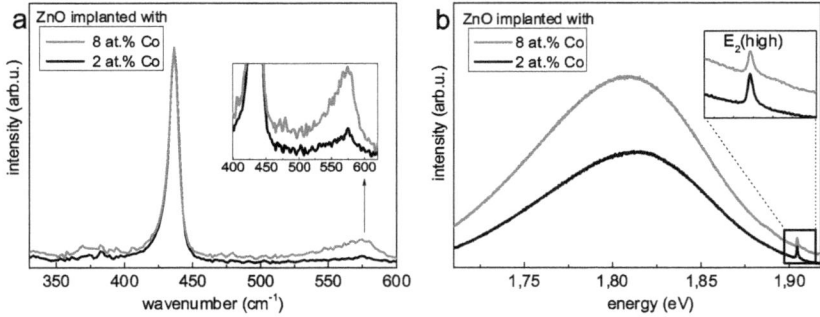

Figure 5.23: *(a) Raman spectra of 2 at.% and 8 at.% Co-implanted ZnO after annealing at 700 °C (excitation 514.5 nm). (b) Corresponding Co^{2+} luminescence in the red spectral range (excitation 632.8 nm).*

5.3   Cobalt-alloyed ZnO

Besides manganese, cobalt is the most studied TM in the context of TM-alloyed ZnO. For this thesis, hydrothermally grown ZnO host crystals were implanted with Co to obtain $Zn_{1-x}Co_xO$ with 2 at.% $\leq$ x $\leq$ 32 at.%. For implantation details, see sections 4.1.1 and 5.1. The Co-implanted series is complemented by polycrystalline, bulk ZnCoO (fabricated via VPT), by MBE-grown ZnCoO layers on sapphire substrate, and by nanocrystalline ZnCoO layers on glass substrate. Table 5.4 gives an overview of all studied ZnCoO samples. In subsection 5.3.1, the results on samples with Co concentrations $\leq$8 at.% are presented. As in the case of Mn, secondary phase formation becomes important for higher TM concentrations, which are discussed in subsection 5.3.2. Subsection 5.3.3, finally, deals with the nanocrystalline layers.

Co-alloyed ZnO	Fabrication	Co concentration
layers	hydrothermally grown ZnO impl. with Co	2-32 at.%
layers	molecular beam epitaxy	0.5-5 at.%
polycrys. bulk	vapor phase transport	< 4 at.%
nanocrys. layers	wet-chemical synthesis	3-12 at.%

Table 5.4: *Overview of the Co-alloyed ZnO samples presented in this thesis.*

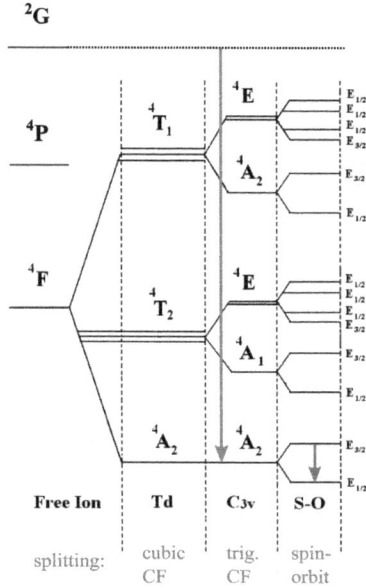

Figure 5.24: *The tetrahedral coordination of substitutional Co^{2+} in ZnO gives rise to crystal field splitting of its 3d levels [Koidl 1977, Kuzian 2006]. This splitting can be described in terms of a cubic part (ideal tetrahedron with $c/a = \sqrt{8/3}$) and a trigonal part (c/a-deviation from the ideal value in the real crystal). The red arrow labels the intra-3d transitions $^2E(G) \rightarrow {}^4A_2(F)$, observed as red emission in Co-alloyed ZnO. The green arrow corresponds to a transition due to the Co^{2+} 3d ground state splitting in ZnO observed at an energy of about 5 cm^{-1} in Raman scattering [Koidl 1977, Szuszkiewicz 2007].*

5.3.1 $Zn_{1-x}Co_xO$ bulk and layers with concentrations ≤ 8 at.%

Hydrothermally grown ZnO was implanted with 2 at.% to 8 at.% Co and studied by Raman scattering, see Figure 5.23a. The A_1(LO) disorder band between 550 cm^{-1} and 600 cm^{-1}, characteristic for TM implantation, is strongly reduced after annealing at 700 °C, indicating substantial healing of the implantation damage. No additional impurity or secondary phase Raman modes occur. Note that the two-shoulder behavior identified as characteristic for Mn-alloyed samples in subsection 5.2.1 is not observed.

When exciting with the 632.8 nm line of a helium-neon laser, most of the Raman signal is covered by a strong luminescence, peaking at about 1.8 eV at room temperature, see Figure 5.23b. This red emission is neither observed in the pure ZnO host crystals nor in ZnO implanted with other

TM ions. The tetrahedral coordination of substitutional Co^{2+} in ZnO gives rise to crystal field splitting of its 3d levels [Koidl 1977, Kuzian 2006]. The crystal field splitting can be described in terms of a cubic part, which corresponds to an ideal tetrahedral coordination with wurtzite lattice constants $c/a = \sqrt{8/3}$, and a trigonal part, which reflects the c/a-deviation from the ideal value in the real crystal, see Figure 5.24. Therefore, the red emission can be identified as intra-3d transitions $^2E(G) \rightarrow {}^4A_2(F)$ of the tetrahedrally coordinated $Co^{2+}(3d^7)$ ion, marked with a red arrow in Figure 5.24. The occurrence of this emission can be used as a proof for substitutional Co^{2+} on Zn sites in wurtzite ZnO. Obviously, the intensity of this emission in the case of the 8 at.% Co-implanted sample is stronger than in the 2 at.% sample. When normalized to the Co concentration, however, the emission is much stronger in the 2 at.% sample. This suggests that the percentage of substitutional Co decreases for higher Co concentrations.

For comparison, $Zn_{1-x}Co_xO$ bulk material was studied. The polycrystalline samples were grown by VPT at a growth temperature of 900-1000 °C and have Co concentrations < 4 at.% [Jouanne 2006]. Room temperature and low temperature Raman spectra of such a bulk ZnCoO sample are shown in Figure 5.25a. In the spectra taken at room temperature, the expected ZnO modes can be observed, i.e. the E_2(high) mode at about 437 cm^{-1}, the E_2(high)-E_2(low) mode at about 330 cm^{-1}, and the disorder band in the LO phonon region. Additionally, the A_1(TO) mode at about 380 cm^{-1} occurs. For the latter, Raman scattering is allowed in this case because of the polycrystalline character of the sample. Its c-axis is not well-defined and therefore the angle between c-axis and laser polarization can not be restricted to 90° as in the experiments on the TM-implanted samples. In the low temperature spectra, not only the difference mode E_2(high)-E_2(low) disappears, but also additional modes occur. The peaks at about 143 cm^{-1} and about 297 cm^{-1} can be identified as CoO magnons [Chou 1976]. The magnon signal of CoO reported for 220 cm^{-1} can not be observed in the experimental configuration used to record these spectra. The temperature behavior of the magnon mode at about 297 cm^{-1} is shown in the inset in Figure 5.25a. Due to stronger correlation between the spins at lower temperatures, it shifts to higher wavenumber values, becomes narrower, and shows higher intensity with decreasing temperature. The formation of CoO identified by these magnon signals can be attributed to the fabrication process because the reported solubility of Co in ZnO usually lies well above the 3-4 at.% of this sample [Jin 2001, Kolesnik 2004]. As these samples were fabricated at high temperature, other Co oxide phases, especially Co_3O_4 and $ZnCo_2O_4$, are not stable during the growth. Note that, due to its rocksalt structure, CoO phonon modes can not be observed as well-defined Raman features. The Raman features of Co oxide modes are discussed in more detail in subsection 5.3.3. An additional Raman feature was detected during experiments on these samples at about 132 cm^{-1}, which could be attributed to a vibration mode of elemental cobalt in hcp structure [Millot 2006, Szuszkiewicz 2007]. Hence, besides the identified CoO, also

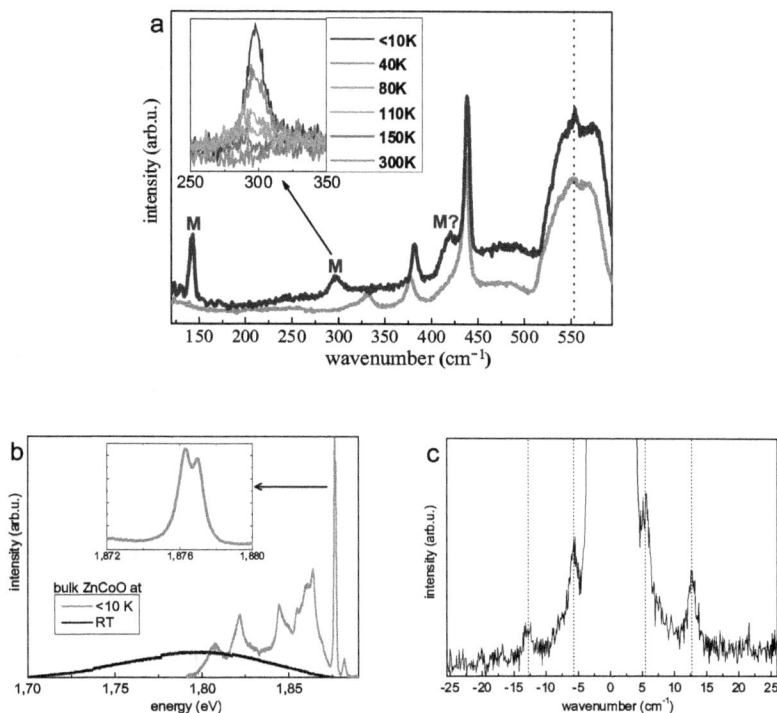

Figure 5.25: *Raman and PL spectra of polycrystalline, bulk $Zn_{>0.96}Co_{<0.04}O$. (a) Additional features in the low temperature Raman spectra caused by scattering from CoO magnons (excitation: $\lambda = 514.5$ nm). (b) Broad red emission due to tetrahedrally coordinated Co^{2+} on Zn sites in ZnO at RT and corresponding luminescence fine structure at low temperature (excitation: $\lambda = 457.9$ nm). (c) Raman scattering signals due to Co^{2+} intra-3d transitions in ZnO ($\sim$5 cm^{-1}) and in CoO ($\sim$13 cm^{-1}) due to 3d ground state splitting (excitation: $\lambda = 488.0$ nm, $T = 15$ K).*

elemental Co is present in these samples already at such low concentrations.

With decreasing temperature, an additional feature evolves at about 420 cm^{-1}, which is not known as a magnon signal of a Co-related alloy. Its temperature behavior suggests a magnon or a localized vibration as origin, but no final assignment can be given here.

Among all TM-alloyed ZnO systems studied for this thesis, the feature at 550 cm^{-1}, as seen in Figure 5.25a, was particularly strong only in Co-alloyed systems. This fact and the temperature behavior suggest an impurity-induced vibration. The B_1(high) silent mode is calculated near this position [Serrano 2007] and activation of silent ZnO phonon modes by impurity incorporation is

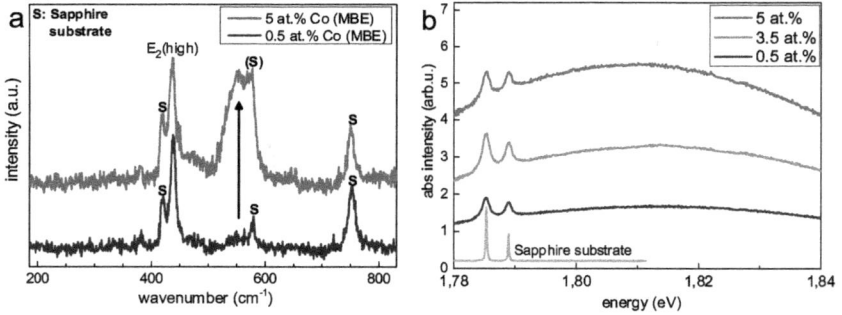

Figure 5.26: *(a) Raman spectra of MBE-grown $Zn_{1-x}Co_xO$ layers with 0.5 at.% and 5 at.% Co (excitation: $\lambda = 514.5$ nm). (b) Red emission of the $Zn_{1-x}Co_xO$ layers due to intra-3d transitions of substitutional Co^{2+} (excitation: $\lambda = 632.8$ nm).*

possible as discussed in [Bundesmann 2003, Manjon 2005]. However, this silent mode is studied in detail in chapter 6 and is found at significantly higher frequency. The resonance behavior of this feature, addressed in subsection 5.3.3, suggests the assignment to a mode with LO symmetry. In the room temperature PL spectra of Co-implanted ZnO, only a broad luminescence band due to tetrahedrally coordinated Co^{2+} occurs in Figure 5.23b. The low temperature spectra in Figure 5.25b reveal additional features. The observed fine structure is due to spin-orbit interaction and phonon satellites of the electronic transitions. Furthermore, interesting features occur at very low wavenumber values, observed for the first time in Raman scattering experiments (Figure 5.25c). The structure at about ± 5 cm^{-1} may correspond to electronic Raman scattering on a transition due to Co^{2+} 3d ground state splitting in $Zn_{1-x}Co_xO$ [Koidl 1977, Szuszkiewicz 2007]. This transition is labeled by a green arrow in Figure 5.24. Another possibility is the Co^{2+} 3d ground state splitting in CoO, which also has an energy corresponding to a Raman shift of 5 cm^{-1} between the lowest level and the second lowest level [Sakurai 1968]. The feature close to ± 13 cm^{-1} can be attributed to Co^{2+} 3d ground state splitting in CoO, where the energy difference between the lowest level and the highest (third) ground state level amounts to 13 cm^{-1} [Sakurai 1968, Szuszkiewicz 2008]. A further, less likely, explanation for the 13 cm^{-1} Raman feature is that the 5 cm^{-1} Co^{2+} transition is shifted if in a locally deformed crystal environment.

The incorporation of Co by ion implantation implies implantation-induced disorder, and in the case of VPT samples, Co oxides were observed already at low concentrations. In contrast, fabrication via molecular beam epitaxy promises more ordered $Zn_{1-x}TM_xO$ systems with weaker secondary phase formation. In the MBE-grown samples studied for this thesis, intrinsic ferromagnetism was found for concentrations <3 at.%, while AFM correlation between the Co^{2+} ions dominates the magnetic behavior for $\geq$3 at.% [Sati 2006]. Like all TM-alloyed ZnO systems of this thesis, these layers are n-type, in this case with residual carrier concentrations $n_e < 10^{18}$ cm^{-3}. In Figures 5.26a and 5.26b, the Raman spectra and red luminescence of the available MBE layers with varying Co concentrations are shown. Due to the high transparency of these thin layers, strong additional signals from the sapphire substrate occur. Interestingly, despite the epitaxial growth, a strong disorder band between 500 cm^{-1} and 600 cm^{-1} is clearly seen in the 5 at.% sample. Like in the bulk samples discussed above, the broad band peaks at 550 cm^{-1}, which was attributed to a impurity-induced mode. The broad Co^{2+} photoluminescence shown in Figure 5.26b confirms the presence of substitutionally incorporated Co ions. However, the luminescence intensity in the 5 at.% samples is only by a factor of 3.2 stronger than in the 0.5 at.% sample, indicating that the fraction of substitutional Co ions decreases with rising Co concentration. This could be related to the magnetic behavior mentioned above: The MBE grown layers show RT FM behavior for low concentrations <3 at.%, but AFM behavior for higher concentrations.

5.3.2 $Zn_{1-x}Co_xO$ layers with concentrations $\geq$16 at.%

For Co-implanted ZnO, the formation of secondary phases was observed only for the sample implanted with 32 at.% Co after annealing at 700 °C. The additional phonon modes occurring in the Raman spectra of this sample clearly indicate the formation of $ZnCo_2O_4$ (Figure 5.27). For comparison, polycrystalline, bulk $ZnCo_2O_4$ was grown by VPT. It shows a structure-rich Raman spectrum with the 5 strongest features peaking at about 487 cm^{-1}, 524 cm^{-1}, 624 cm^{-1}, 692 cm^{-1}, and 709 cm^{-1}. These wavenumber values are in excellent agreement to the positions of the additional modes in the 32 at.% Co-implanted sample (Figure 5.28). No other secondary phases were observed for the Co-implanted samples by Raman scattering. The formation of CoO cannot be excluded from these room temperature spectra because it exhibits rock-salt structure in stoichiometric bulk form, which does not posses Raman-active phonon modes. However, formation of $ZnCo_2O_4$ is favored under annealing in air and decomposition of Co_3O_4 to CoO requires temperatures between 800 °C and 900 °C [Schumm 2007, Thakur 2007]. The $ZnCo_2O_4$ polycrystalline sample was also studied with respect to its luminescence properties. In contrast to all Co-alloyed ZnO samples, no luminescence was detected in the red spectral region, in which the luminescence caused by tetrahedrally coordinated Co^{2+} is located.

Besides the 0002 and 0004 Bragg peaks of ZnO, also a third diffraction peak in Figure 5.29 is not related to the implanted Co. The unidentified signal at about 52.8° (marked with an asterisk)

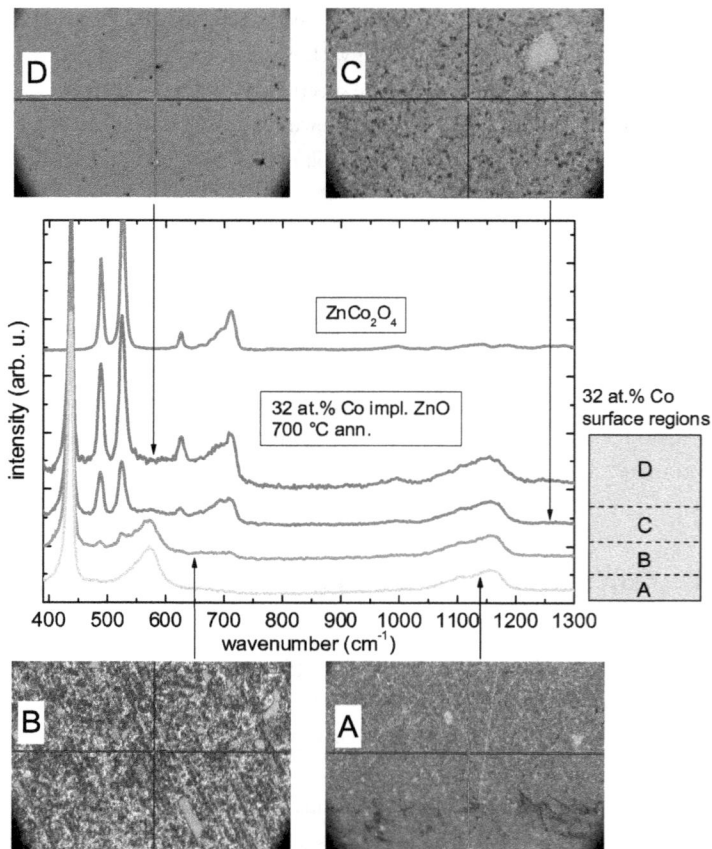

Figure 5.27: *Raman spectra of different spots on the 32 at.% Co-implanted ZnO sample after 700 °C annealing (excitation: λ = 514.5 nm). The optical microscope pictures show the studied surface regions A, B, C, D of this sample with inhomogeneously distributed oxide formation. Raman spectra from bottom to top: laser focused in the middle of the surface region (A), (B), (C), and (D).*

was observed in several samples with different TM species and varying TM concentrations. It is attributed to the quasi-forbidden 0003 reflex of ZnO. Despite the clear identification of $ZnCo_2O_4$ by strong Raman signals, no diffraction peaks in Figure 5.29 can unambiguously be assigned to this secondary phase. This is surprising although the sensitivity advantage of Raman in

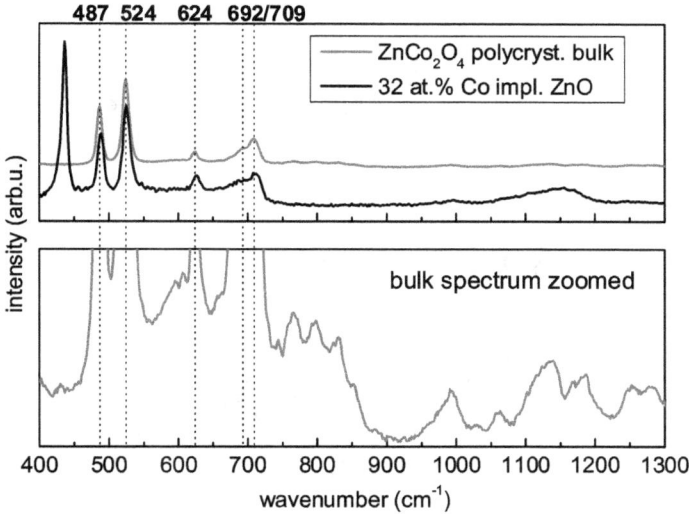

Figure 5.28: *Raman spectra of polycrystalline, bulk $ZnCo_2O_4$ and of 32 at.% Co-implanted ZnO (spot D) after 700 °C annealing (excitation: $\lambda = 514.5$ nm).*

comparison to XRD was reported before for Co oxide formation in Co-alloyed ZnO [Wang 2007]. Two additional peaks in the diffractogram in Figure 5.29 are attributed to secondary phases. The signal at about 43.4° could be due to hcp zinc, which has its strong 10$\bar{1}$1 Bragg peak at about 43.23° [JCPDS 1997]. The formation of elemental zinc could be enabled by the structural impact of the implantation procedure. The additional peak at about 44.3° is assigned to the 0002 peak of hcp Co at 44.76° [JCPDS 1997]. The 111 peak of cubic Co lies near this diffraction peak, too. Though, the usual crystal structure of elemental Co is hexagonal, and furthermore, hcp Co was observed before in Co-alloyed ZnO by synchrotron XRD and by Raman scattering [Millot 2006, Potzger 2008, Szuszkiewicz 2007, Zhou 2008/1]. The Raman feature of hcp Co, however, lies in a spectral region not accessible in these Raman experiments. While the identified hcp Zn and hcp Co are plausible secondary phases, their assignment is tentatively because corresponding higher-angle reflections were not accessible in the recorded angle range.

The oxide phase is inhomogeneously distributed over the sample surface: A gradual increase of the Co oxide concentration from one side of the sample to the other can be deduced from the Raman spectra taken in different regions (A, B, C, D in Figure 5.27). These regions also show very different surface qualities under the microscope after the 700 °C annealing. Inhomogeneous

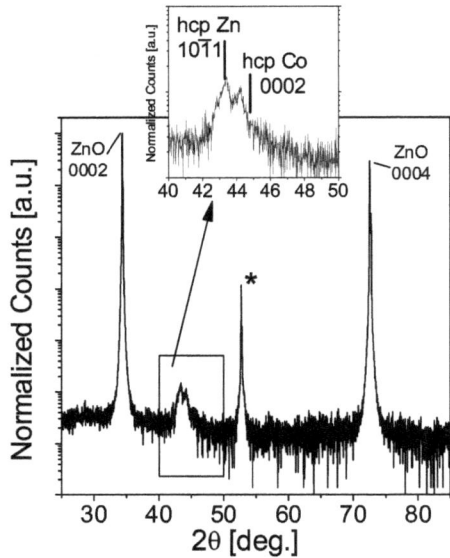

Figure 5.29: *XRD diffractogram of the 32 at.% Co-implanted ZnO sample. Besides the 0002 and 0004 Bragg peaks of the ZnO host, three additional features can be observed. The peak marked with an asterisk corresponds to the quasi-forbidden 0003 reflex of ZnO. The other additional features are ascribed to the $10\bar{1}1$ Bragg peak of elemental hcp Zn and the 0002 Bragg peak of elemental hcp Co, reported in [JCPDS 1997].*

implantation is improbable as the surface was completely homogeneous after the implantation, but can not be excluded. Another cause for the inhomogeneous oxide distribution could be a temperature gradient during annealing or during the cooling down of the sample.

The observed intensity increase of the oxide phase modes from A to D is accompanied by a decrease of the disorder band in the $A_1(LO)$ region. Region D shows the strongest oxide phase Raman modes, nearly identical to the bulk $ZnCo_2O_4$, but also the smallest disorder band and the most homogeneous surface under the microscope. Accordingly, the intensity of the Co^{2+} luminescence signal at about 1.8 eV is considerably enhanced in the sample surface region with reduced Co oxide concentration. These Raman scattering and luminescence observations can again be attributed to a self purification process induced by the thermal treatment: During annealing, Co ions migrate to the surface and form an ordered, partially closed Co oxide layer, accompanied by an improved crystal quality of the nearby ZnO. This layer is (almost) closed in the region D, which consequently shows the most homogeneous surface under the microscope and the strongest secondary phase phonon modes in the Raman spectra.

5.3.3   Nanocrystalline ZnO:Co layers

ZnCoO thin layers were prepared by sol-gel dip-coating. The resulting layers consist of nanocrystals with an average size of about 200 nm and Co concentrations of 3 at.% to 12 at.%. For more
details on the fabrication see [Bhatti 2007, Zhou 2007/1]. The samples were studied by micro-
Raman measurements with the standard Ar^+ laser lines of the Dilor XY setup (457 nm $\leq \lambda \leq$
514 nm).

Figure 5.30a shows the Raman spectra of the $Zn_{1-x}Co_xO$ layers for different Co concentrations
(3 at.%, 7 at.%, 12 at.%) and different annealing steps. The well-known ZnO Raman modes can
be seen in all spectra: E_2(high) at about 437 cm^{-1}, A_1(TO) at about 380 cm^{-1}, and E_2(high)-
E_2(low) at about 330 cm^{-1}. Furthermore, the broad disorder band appears in the spectral region
between 500 cm^{-1} to 600 cm^{-1}. The disorder-related character of this band is once more confirmed by resonance experiments as shown in Figure 5.31a. The structure is largely enhanced by
exciting with a laser line of higher energy. A similar enhancement was discussed in subsection
5.2.1 for ZnMnO and could be attributed to impurity-induced Fröhlich scattering. Relative to
the non-polar E_2(high) mode, the enhancement is particularly strong at about 570 cm^{-1}. In this
region, the polar LO modes E_1 and A_1 (570-585 cm^{-1}) are situated, which are predominantly
affected by Fröhlich scattering resonance. Furthermore, also the resonance enhancement of the
550 cm^{-1} feature is strong. As possible origin of this feature an impurity-induced mode was
suggested in subsection 5.3.1. The observed resonance behavior indicates that this additional
mode possesses LO character.

Additional features appear in the spectra of Figure 5.30a, which can be explained as vibrations of Co-related secondary phases. Note that in Figure 5.31a only the ZnO modes in the
LO region show resonance enhancement, but not these additional modes. The peaks at about
488 cm^{-1}, 527 cm^{-1}, and 714 cm^{-1} can be assigned to non-stoichiometric Co_3O_4 [Hadjiev 1988].
There are also broad structures between 650 cm^{-1} and 720 cm^{-1} and around 460 cm^{-1}, which can
both be ascribed to CoO [Gallant 2006, Zhang 2006, Zhou 2007/1]. Furthermore, also $ZnCo_2O_4$
formation is possible, which shows a similar broad structure in Raman spectra from 650 cm^{-1}
to 720 cm^{-1} [Samanta 2006].

A two-step annealing procedure was chosen to study the formation of precipitates during such
after-growth treatments. Both steps (700 °C, 900 °C) have temperatures sufficiently high for
pronounced Co oxidation. $T_{ann} = 900$ °C was chosen in order to provoke the migration of interstitial Co, resulting in an improved quality of the ZnCoO. At the same time, the risk of Co
oxidation on the sample surface increases. Before the first annealing step, most of the layers
did not show any ZnO- or Co-related Raman signals due to poor crystal quality. This changed
with the first annealing at 700 °C for 15 min in air. In the second step, the sample was slowly

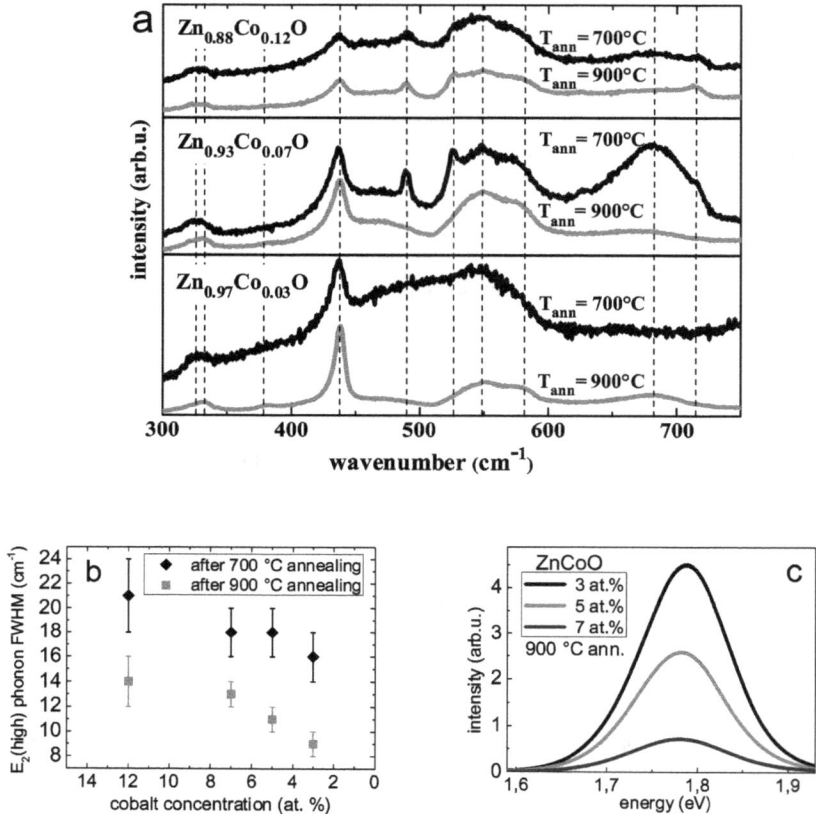

Figure 5.30: *(a) Raman spectra of $Zn_{1-x}Co_xO$ with $x = 3$ at.%, 7 at.%, and 12 at.%, recorded after annealing at $700\,^\circ C$ and $900\,^\circ C$ (excitation: $\lambda = 514.5$ nm). (b) Full width at half maximum of the E_2(high) phonon mode versus Co concentration after $700\,^\circ C$ and $900\,^\circ C$ annealing. The disorder-induced broadening is reduced by the $900\,^\circ C$ annealing and scales with the Co concentration. (c) Co^{2+} luminescence of the nanocrystalline samples after $900\,^\circ C$ annealing (excitation: $\lambda = 632.8$ nm).*

heated to 900 °C in air over a period of 4 hours, then the temperature was held constant for 15 min. Afterwards, the sample was slowly cooled down to room temperature. The spectra of the sample with 3 at.% Co show that after the 700 °C annealing, the crystal structure of the ZnO is still relatively bad, leading to a broad E_2(high) mode and a strong LO disorder band at

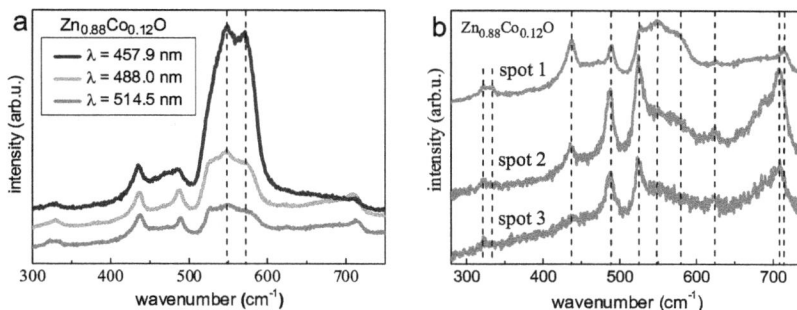

Figure 5.31: *Raman spectra of the $Zn_{0.88}Co_{0.12}O$ layer: (a) resonance behavior studied by varying excitation wavelengths (457.9 nm $\leq \lambda \leq$ 514.5 nm), (b) micro-Raman scattering analysis of different sample surface spots (excitation: λ = 514.5 nm).*

500-600 cm^{-1}. Smaller peaks and strongly reduced background noise in the Raman spectra after the 900 °C annealing indicate a significant improvement of the crystalline quality. For example, the width of the E_2(high) phonon mode is reduced by up to 40% due to the 900 °C thermal treatment for all concentrations, see Figure 5.30b. The scaling of the E_2(high) FWHM with the Co concentration can be explained by a stronger disorder and alloy potential fluctuations, as discussed in subsection 5.2.3 for ZnMnO nanoparticles.

There is no formation of Co_3O_4 observed for the 3 at.% sample. The structure at 650 cm^{-1} to 720 cm^{-1}, however, may be assigned to $ZnCo_2O_4$, as described before. In contrast, for the layer with 7 at.% Co, Co_3O_4 vibrations can be observed after the 700 °C annealing. Additionally, the CoO- or $ZnCo_2O_4$-caused broad structure is visible. After the 900 °C annealing, the signal of Co_3O_4 has disappeared and only the broad CoO / $ZnCo_2O_4$ structure occurs, but strongly reduced in intensity. For the layer with the highest Co concentration (12 at.%), the crystal quality after the 700 °C annealing is the least perfect and Co_3O_4 modes are strong both after the 700 °C and the 900 °C annealing.

To explain these results, the formation and degradation temperatures of the participating oxides must be taken into account. The formation of cobalt oxides is possible for temperatures well below 700 °C. Therefore, Co_3O_4, CoO, and $ZnCo_2O_4$ can be produced during the annealing procedure. Moreover, the formation of Co oxides is favored compared to metal Co clusters for annealing in air. When heating the sample, the Co available on the surface and Co migrating to the surface of the nanocrystals due to the temperature effect are oxidized. This Co oxide formation can be observed after the first annealing step for the samples with concentrations >3 at.%. In the second annealing step, the temperature is higher than 700 °C for some hours before reaching 900 °C, so most of the available Co is transported to the surface and oxidized. A temperature of 900 °C is higher than the reduction temperature of Co_3O_4 into CoO [O'Neill 1985]. Therefore, the disappearance of the Co_3O_4 vibration modes after the second annealing process for the sample with 7 at.% Co corresponds to Co_3O_4 reduction. For this layer, all available Co was oxidized as well as all formed Co_3O_4 was reduced to CoO. That means, there is no Co available for the formation of new Co_3O_4 during the cooling process, when the temperature is within the Co_3O_4 formation range for several hours. The strong decline in intensity of the 650 cm^{-1} to 720 cm^{-1} band during the 900 °C annealing could be due to the reduction of $ZnCo_2O_4$. The origin of this band (CoO or $ZnCo_2O_4$) is addressed in more detail later. The layer with 12 at.% Co still shows formation of Co_3O_4 in the cooling period after the second annealing: The migration process of Co to the nanocrystalline surface is still not completed after the two annealing steps for this sample with the highest Co concentration. These findings concerning the formation and reduction of cobalt oxides on the surface of $Zn_{1-x}Co_xO$ layers are in agreement with XRD studies on similar ZnCoO nanocrystals [Bhatti 2007]. The micro-Raman method gives the opportunity to analyze surfaces with spatial resolution, though in this case with a laser spot of about 1 μm there is an averaging over an ensemble of nanocrystallites. The strongest surface inhomogeneity in the corresponding Raman spectra was seen in the 12 at.% Co sample after the 900 °C annealing. As Figure 5.31b shows, the Co oxide concentration varies for different spots (1, 2, 3) on the sample. Interestingly, the broad signal at about 650 cm^{-1} to 720 cm^{-1} shifts between spot 1, where the ZnO modes are still strong, and spots 2 and 3, where the Co oxide signals dominate. This could reflect the transition of the Co oxide vibrations from $ZnCo_2O_4$ to CoO character. The shift is in accordance with literature data for CoO and $ZnCo_2O_4$ [Gallant 2006, Samanta 2006].

The formation of Co-related secondary phases has an important impact on the magnetic properties of ZnCoO, as CoO and Co_3O_4 are anti-ferromagnetic or paramagnetic, and metal Co is ferromagnetic. The magnetic properties of the samples were determined by SQUID measurements [Zhou 2007/1]. While the 3 at.% and 5 at.% samples showed paramagnetic signals at room temperature, a weak ferromagnetic signal was observed for the 7 at.% and 12 at.% samples after 700 °C annealing. The latter may be attributed to elemental Co clusters dominating the magnetic properties of the layers with high concentrations although they could neither be identified by Raman scattering nor by XRD experiments. Still, nanocrystalline precipitates of elemental Co could be below the sensitivity of both Raman scattering and conventional XRD.

Compared to the implanted layer discussed above, the secondary phase formation occurs for much lower concentrations in these nanocrystalline layers ($x_{nano} \geq 5$ at.% compared to $x_{impl} \geq 32$ at.%). In similar $Zn_{1-x}Co_xO$ nanocomposites prepared by a sol-gel method and studied by XRD and XAFS (X-ray absorption fine structure spectroscopy), mainly substitutional Co incorporation was found below 5 at.%, but Co_3O_4 precipitation started at x $\geq$ 10 at.% [Shi 2007]. This lower solubility limit of Co in ZnO nanostructures compared to bulk ZnO can be attributed to the strong self-purification tendency of such systems due to their reduced dimensionality and large surface-to-volume ratio [Dalpian 2006]. The migration process of Co ions to the surface, especially during annealing, is energetically favored and its subsequent oxidation leads to more pronounced secondary phase formation in nanocrystalline material.

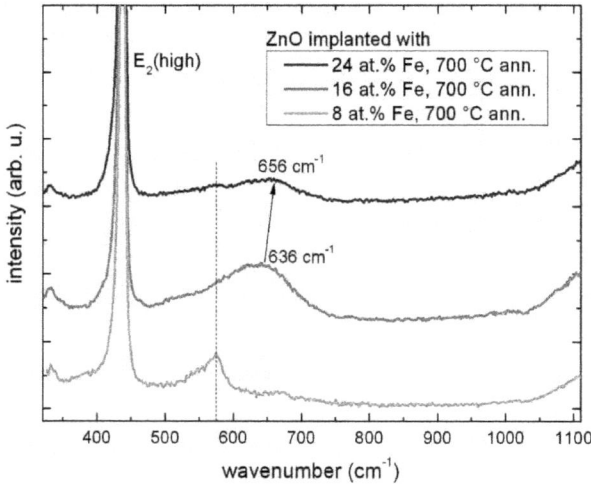

Figure 5.32: *Raman spectra of Fe-implanted ZnO samples of different concentrations after 700 °C annealing indicate precipitate formation for concentrations > 8 at.% (excitation: $\lambda = 514.5$ nm).*

5.4 Iron-alloyed ZnO

Fe-alloyed ZnO	Fabrication	Fe concentration
layers	hydrothermally grown ZnO impl. with Fe	2-32 at.%

Table 5.5: *Overview of the Fe-alloyed ZnO samples presented in this thesis.*

Hydrothermally grown ZnO crystals were implanted with Fe concentrations from 2 at.% up to 32 at.%. For 8 at.% Fe and after 700 °C annealing in air, the Raman spectrum is shown in Figure 5.32. The dominating E_2(high) mode and the weak A_1(LO) disorder band between 550 cm^{-1} and 600 cm^{-1} indicate that most implantation damage was healed by the thermal treatment. No secondary phases or additional modes are visible. This is in accordance with the results for Mn-, Co-, and Ni-implanted ZnO of low TM concentrations.

To study the influence of the annealing environment on the secondary phase formation, both

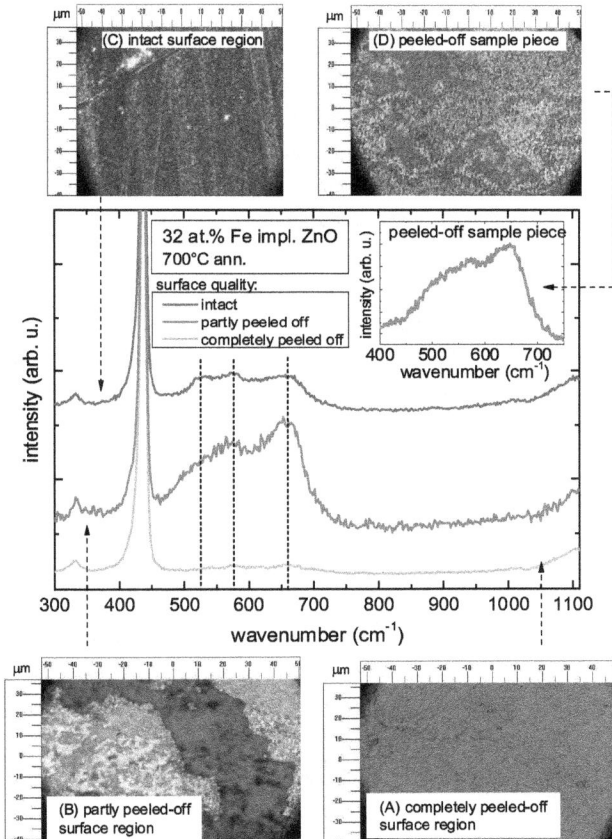

Figure 5.33: *Raman spectra of different spots on the 32 at.% Fe-implanted ZnO sample after 700 °C annealing (excitation: λ = 514.5 nm). The optical microscope pictures show strong inhomogeneities in the studied surface regions. Raman spectra from bottom to top: laser focused on (A) completely peeled-off surface region, (B) partly peeled-off surface region, (C) intact surface region, and (D) peeled-off sample piece (inset).*

annealing in air and vacuum annealing were conducted for the samples with high Fe concentrations. For annealing in air, secondary phases were detected after thermal treatment with 700 °C for all Fe-implanted samples with concentrations $\geq$16 at.% (Figures 5.32 and 5.33). In contrast to the Mn- and Co-implanted samples, however, the observed broad and unstructured

Raman bands complicate a clear identification. Within a broad band ranging from 500 cm^{-1} to 700 cm^{-1}, features occur which were not observed for the samples with Fe concentrations <16 at.% nor for annealing <700 °C. Their frequencies lie in the spectral region of strong Fe_3O_4 and FeO modes [Chourpa 2005, De Faria 1997]. However, no clear Raman signature is visible, which indicates the absence of well-ordered, stoichiometric Fe oxides.

In detailed lateral mapping, already the Raman spectra of the samples implanted with 16 at.% and 24 at.% Fe indicate secondary phase formation with a broad Raman band between 600 cm^{-1} and 700 cm^{-1} (Figure 5.32). Nevertheless, they still exhibit a nearly homogeneous surface with only few minor inclusions. These inclusions turned out to be very sensitive to optical irradiation. This poor thermal stability implies that part of the phase formation took place during the cooling down of the sample after annealing. For the sample implanted with 32 at.%, the high implantation dose combined with 700 °C thermal treatment lead to destruction of large parts of the surface. After the annealing, the surface is characterized by intact regions as well as partly or completely destroyed regions due to peeled-off surface pieces (Figure 5.33). Raman spectra taken in regions where the surface was completely peeled off correspond to pure ZnO. In the spectra of partly destroyed surface regions, a broad feature occurs, which may correspond to non-stoichiometric Fe oxides as discussed above, and is also observed for a peeled-off surface piece (see inset in Figure 5.33).

The ZnO 0002 (34.4°) and 0004 (72.6°) reflections as well as the quasi-forbidden 0003 peak at about 52.8° are observed in the diffractogram in Figure 5.34a. An additional peak occurs at about 56.8° which presumably corresponds to the strong 511 Bragg peak (56.63°) of $ZnFe_2O_4$ [JCPDS 1997]. From XRD, an average lattice constant of a = 8.4 Å is deduced for the secondary phases. Assuming the formation of oriented spinel structures on ZnO, the resulting lattice deviation of about 6% may explain the described surface effects, not observed for samples implanted with Mn, Co, or Ni ions. With the Scherrer equation [Scherrer 1918], an estimated size of about 12 nm was calculated for the segregations. The corresponding higher-angle peaks lie outside the accessible angle range of the used XRD setup. However, zinc ferrite precipitates have been reported as secondary phase in Fe-alloyed ZnO [Zhou 2007/2].

As shown in subsection 5.2.1, already the ZnO host crystals show luminescence features attributed to residual Fe ions. For the highly Fe-implanted ZnO samples, an additional broad luminescence could be identified between the green and the red spectral range (Figure 5.34b). The 514.5 nm line of the Ar ion laser was used for excitation, and therefore, the fine structure in the spectra between 2.25 eV and 2.4 eV corresponds to the phonon Raman signals of the implanted systems. In contrast, the broad luminescence between 1.8 and 2.2 eV is not present in the pure ZnO crystal and therefore related to the implantation. As both features are especially strong for the 16 at.% Fe concentration, they are not related to the identified secondary phases

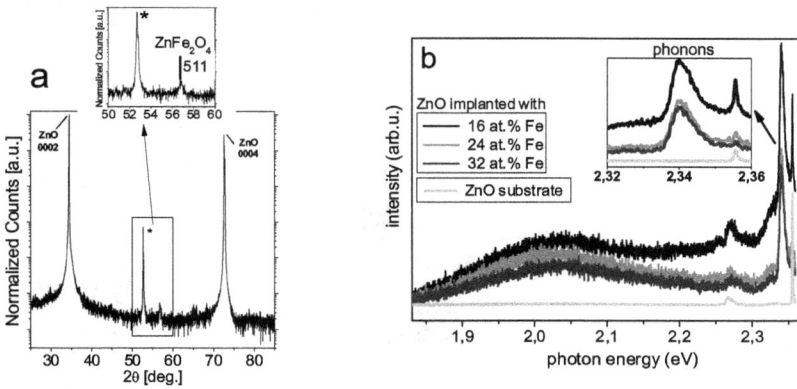

Figure 5.34: *(a) XRD diffractogram of 32 at.% Fe-implanted ZnO. Besides the 0002 and 0004 Bragg peaks of the ZnO host, two additional features are observed. The peak marked with an asterisk corresponds to the quasi-forbidden 0003 reflex of ZnO. The other additional feature is assigned to the 511 Bragg peak of $ZnFe_2O_4$, reported in [JCPDS 1997]. (b) Photoluminescence spectra of 16 at.%, 24 at.%, and 32 at.% Fe-implanted ZnO as well as pure ZnO substrate for comparison (excitation: $\lambda = 514.5$ nm). Besides the Raman features near the excitation energy of 2.41 eV, a broad luminescence is observed in the implanted systems, which can be identified as the green defect or impurity band of ZnO.*

which are stronger for the higher concentrations. Hence, it is attributed to the green luminescence of ZnO, which is reported to be due to either intrinsic ZnO defects or impurities in ZnO (subsection 3.1.3).

As for the highly Mn-implanted ZnO crystals, additional HRTEM and EDX line scan experiments were conducted for the 32 at.% Fe-implanted ZnO crystal after 900 °C annealing in air. The experimental details of these methods were described in subsection 5.2.1. Figure 5.35 shows TEM pictures of secondary phase clusters, which are again elongated and aligned to the subjacent ZnO host crystal and have sub-μ size. The strong variation of the contrast is due to structural disorder of the implanted ZnO. As in the case of the 32 at.% Mn-implanted samples, EDX line scans confirm a substantially higher TM concentration within the secondary phase clusters.

The identified, strong TM oxide formation is expected to be at least partly caused by the presence of oxygen during annealing in air. To verify this effect, also high vacuum annealing experiments were performed with Fe-implanted ZnO samples (2 at.% - 32 at.%) for comparison. The samples were fabricated exactly as the samples described above. In addition, also the vacuum annealing

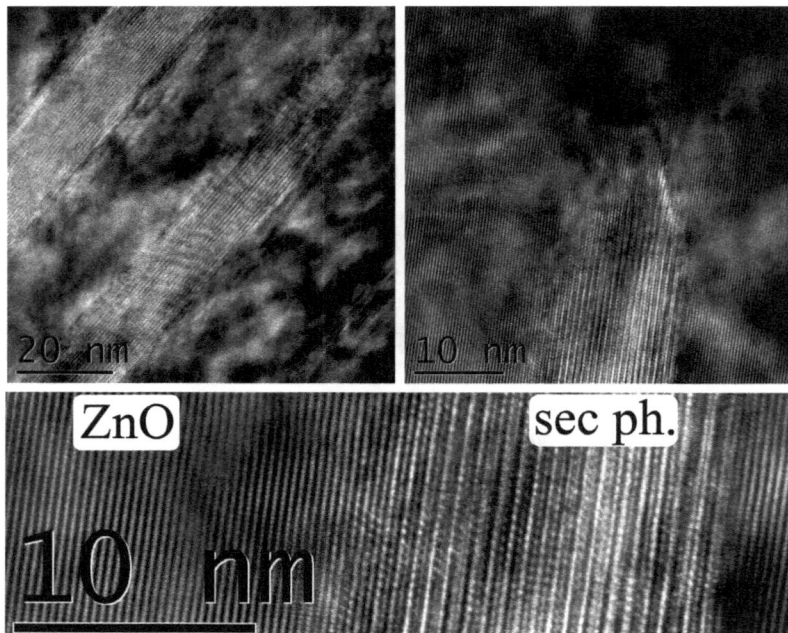

Figure 5.35: *High-resolution TEM pictures of secondary phase clusters on the surface of the 32 at.% Fe-implanted sample after 900 °C annealing in air.*

was performed exactly as the annealing in air, i.e. with a duration of 30 min and a temperature of 700 °C. Already under the microscope, a significantly better surface quality is observed, even for the 32 at.% implanted sample (inset in Figure 5.36). This is in contrast to the partial surface destruction after annealing in air (Figure 5.33). Furthermore, the Raman spectra indicate a reduced secondary phase segregation in Figure 5.36. From pure ZnO to 24 at.% Fe-implanted ZnO, only slight changes are visible in the Raman spectra of the vacuum-annealed samples. Only the 32 at.% Fe-implanted sample shows Raman features similar to the secondary phase signals discussed above. Still, their intensity is weaker than in the spectrum of the 32 at.% air-annealed sample. In summary, the vacuum annealing proves to be the method with better resulting surface quality and less pronounced secondary phase segregation, but the formation of Fe-related oxides is still not completely suppressed for high concentrations.

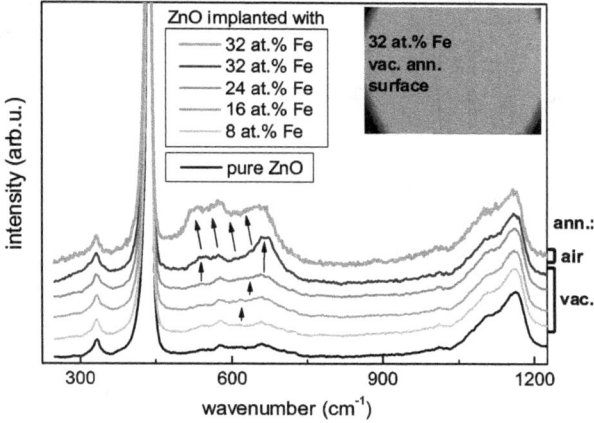

Figure 5.36: *Raman spectra of Fe-implanted ZnO crystals with concentrations between 8 at.% and 32 at.% after vacuum annealing (excitation: λ = 514.5 nm). For comparison, the spectrum of 32 at.% Fe-implanted ZnO annealed in air is shown. The inset shows the homogeneous surface of the vacuum-annealed 32 at.% sample.*

5.5 Nickel-alloyed ZnO

Ni-alloyed ZnO	Fabrication	Ni concentration
layers	hydrothermally grown ZnO impl. with Ni	2-32 at.%

Table 5.6: *Overview of the Ni-alloyed ZnO samples presented in this thesis.*

Ni-alloyed ZnO was studied in the form of implanted samples with Ni concentrations between 2 at.% and 32 at.%. For the samples implanted with ≤8 at.%, no secondary phases could be identified by Raman scattering in Figure 5.37a. The corresponding spectra show the ZnO E_2(high) and E_2(high)-E_2(low) modes, and additionally the A_1(LO) disorder band. The intensity of the disorder band scales with the Ni concentration.

Also for the Ni-implanted samples with higher concentrations, no secondary phases could be detected by the means of Raman spectroscopy even for 32 at.% and after annealing at 700 °C, see Figure 5.37b. The intensity of the A_1(LO) disorder band shows a substantial decrease with increasing annealing temperatures, reflecting the improved crystalline quality of the ZnO host crystal upon annealing. With optical microscopy (50x), an intact surface is found for all

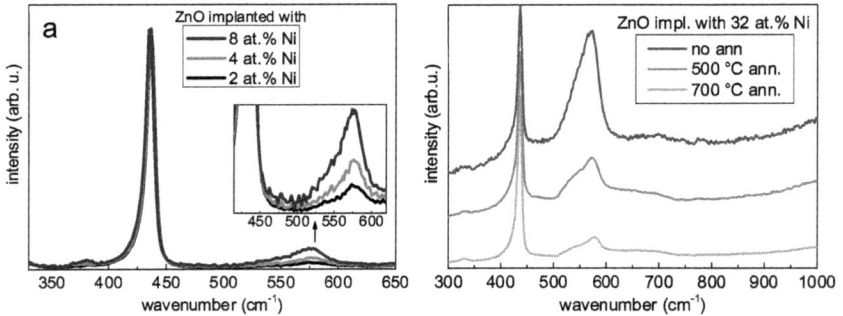

Figure 5.37: *(a) Raman spectra of 2 at.%, 4 at.%, and 8 at.% Ni-implanted ZnO after 700 °C annealing, intensity normalized to the E_2(high) mode (excitation: $\lambda = 514.5$ nm). The intensity of the $A_1(LO)$ disorder band scales with the Ni concentration. (b) Raman spectra of 32 at.% Ni-implanted ZnO after different annealing steps show increasing healing effects with increased temperatures but no precipitate formation even after 700 °C (Note the XRD results of the sample!). Spectra from top to bottom: no annealing, after annealing at 500 °C, and after annealing at 700 °C (excitation: $\lambda = 514.5$ nm).*

concentrations and annealing treatments and no inclusions are resolved on the surface.

In contrast, additional features occur in the XRD measurements of all Ni-implanted samples with Ni concentrations ≥ 16 at.% at about 37.1°, 44.4°, and 79.1°, which can be assigned to NiO and elemental Ni. The strong 111 and 222 Bragg peaks of NiO are reported at 37.25° and 79.41° [JCPDS 1997], respectively, corresponding well to the observed peaks at about 37.1° and 79.1°. The third additional feature at 44.4° is attributed to the strong 111 Bragg reflection of cubic Ni at 44.51° [JCPDS 1997]. The 01$\bar{1}$1 Bragg peak of hcp Ni also lies near this position, but elemental Ni crystallizes in cubic form in ambient conditions. Moreover, the formation of cubic Ni in Ni-alloyed ZnO was reported before [Potzger 2008, Zhou 2008/1]. The grain size of the Ni and NiO precipitates in the samples studied for this thesis could be estimated to about 5 nm and 9 nm, respectively [Scherrer 1918]. The identified secondary phases are in accordance with the absence of secondary phase peaks in the Raman spectra. For the cubic phases Ni and NiO in their stoichiometric form, no Raman-active phonon modes are allowed. Raman scattering from NiO magnons can only be observed in low temperature experiments [Dietz 1971].

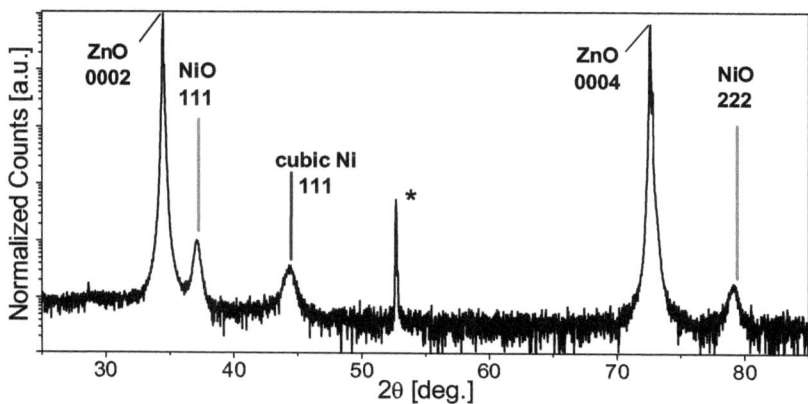

Figure 5.38: *XRD diffractogram of the 32 at.% Ni-implanted ZnO sample. Besides the 0002 and 0004 Bragg peaks of the ZnO host, four additional features can be observed. The peak marked with an asterisk corresponds to the quasi-forbidden 0003 reflex of ZnO. The other additional features correspond to the 111 and 222 Bragg peaks of NiO and the 111 peak of elemental cubic Ni, reported in [JPCDS 1997].*

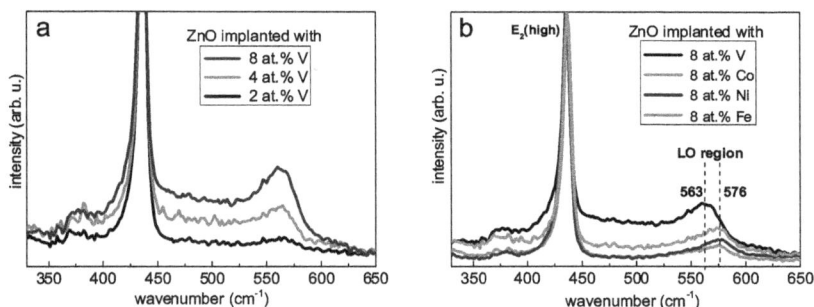

Figure 5.39: *(a) Raman spectra of 2 at.%, 4 at.%, and 8 at.% V-implanted ZnO after 700 °C annealing, normalized to the E_2(high) mode (excitation: $\lambda = 514.5$ nm). (b) Raman spectra of 8 at.% V-, Fe-, Ni-, and Co-implanted ZnO after 700 °C annealing, intensity normalized to the E_2(high) mode (excitation: $\lambda = 514.5$ nm).*

V-alloyed ZnO	Fabrication	V concentration
layers	hydrothermally grown ZnO impl. with V	2-8 at.%

Table 5.7: *Overview of the V-alloyed ZnO samples presented in this thesis.*

5.6 Vanadium-alloyed ZnO

For this thesis, vanadium-implanted ZnO crystals with V concentrations ≤ 8 at.% were studied (Table 5.7). No secondary phases or additional modes could be identified in the Raman spectra of these samples. Once more, the most prominent features in Figure 5.39a are the ZnO E_2(high) mode and the broad disorder band between 500 cm^{-1} and 600 cm^{-1}. Nevertheless, there are differences to the Raman spectra of the systems implanted with other TM ions. Figure 5.39b shows the Raman spectra of 8 at.% V-implanted ZnO compared to 8 at.% Fe-, Ni-, and Co-implanted ZnO. Obviously, the maximum in the A_1(LO) region is red-shifted by about 10 cm^{-1} in the case of V. Such a shift was already identified and discussed in the case of unannealed, heavily implanted ZnO (section 5.1) and was attributed to reduced symmetry due to structural defects. The V-implanted ZnO systems were additionally studied in a depth profile series (Figure 5.40). The method is described in subsection 5.2.1. With the focus being deep within the sample (-22.5 μm), the Raman spectrum of the pure ZnO substrate is observed. When it is moved from this position towards the surface of the sample (0 μm), a strong background occurs, reflecting the phonon DOS [Serrano 2004]. This background is much stronger than in the case of the other TM species (Figure 5.39b) and rises with the V concentration (Figure 5.39a). This also indicates an especially strong crystal disorder due to the V incorporation. Since V has the lowest mass of the studied TM ions, this can not be due to the implantation damage. On the other hand, the V mass shows the biggest difference of all studied TM ions with respect to the substituted Zn atoms ($m_V \approx 0.78\, m_{Zn}$). Hence the observed Raman scattering properties, namely a strong background as well as a 10 cm^{-1} shift of the A_1(LO) signal, can be attributed to the structural disorder caused by the different cation masses of V and Zn.

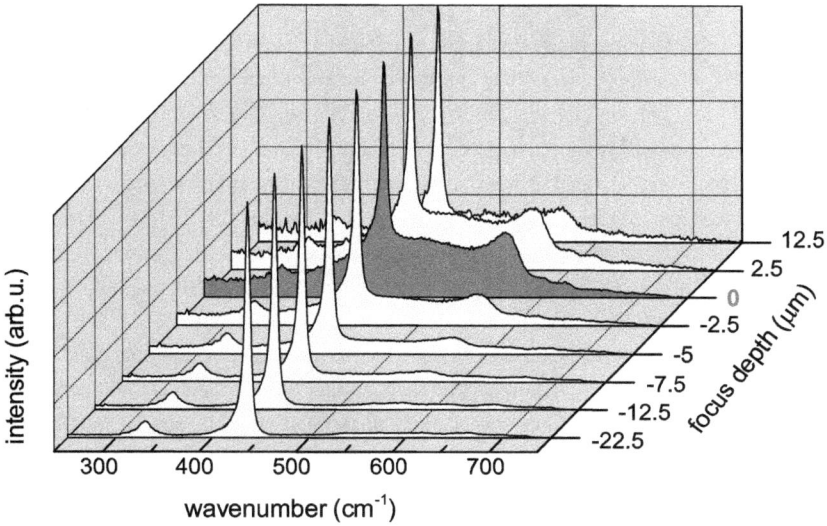

Figure 5.40: *Series of Raman experiments for 8 at.% V-implanted ZnO using different focus depths. Negative values correspond to a focus within the sample, zero to the sample surface, and focus positions above the sample surface are denoted with positive values. The spectra were taken after 700 °C thermal annealing and are normalized to the E_2(high) mode (excitation: $\lambda = 514.5$ nm).*

5.7 Conclusion

TM-alloyed ZnO bulk, layer, and nanocrystalline systems with varying TM concentrations and fabricated by different growth processes were studied by Raman scattering and complementary methods. In this section, the results are evaluated with respect to the potential applications of such systems (subsection 3.2.2).

For all studied TM species, implanted layers were available, which allow controllable TM concentration and distribution within the implanted layers. On the other hand, crystal disorder induced by the implantation is inevitable, and thermal treatment becomes necessary. At low concentrations ≤8 at.%, most implantation damage could be healed with an annealing temperature of 700 °C, and no secondary phase formation could be observed. Hence, samples without precipitates could be realized with lower TM concentrations ≤8 at.%, where also the implantation damage is comparatively low and annealing temperatures <700 °C are sufficient. Another technological improvement could be the use of vacuum annealing. In high vacuum annealing experiments with 32 at.% Fe-implanted ZnO, systems with a significantly better surface quality and less pronounced secondary phase formation were achieved. However, the formation of Fe-related precipitates was still not completely suppressed for this concentration, even after vacuum annealing.

Clear results were achieved for the ZnO crystals implanted with TM concentrations of 16 at.% to 32 at.% after annealing in air. While even for the highest implantation doses the ZnO surface retained its wurtzite character, formation of elemental TM and TM oxides occurred for all studied TM species after the thermal treatment. Some identified and other potential secondary phases are listed in Table 5.8 together with their magnetic properties. Already small and local inclusions can strongly influence the magnetic properties of such systems. The local distribution of the phase formation on the surface strongly depends on the TM type and concentration. For example, the largest studied concentration of 32 at.% results in a surface peel off in the case of Fe, but in a homogeneous distribution of secondary phase islands in the μm range for Mn. In several samples with small local precipitates, no secondary phases could be detected by XRD. On the other hand, Raman spectroscopy fails at the identification, for example, of elemental Ni clusters, which were detected by XRD. Hence, a combination of both methods proves to be necessary for the analysis of secondary phase formation in TM-implanted ZnO.

Promising systems were the studied Co-alloyed MBE layers. Ferromagnetism at room temperature was reported for low concentrations and no secondary phases could be identified by Raman scattering and XRD. However, the existence of elemental Co clusters or CoO inclusions can not be excluded. Moreover, it should be noted that ferromagnetic behavior was found in Co-implanted ZnO layers with concentrations of 3-5 at.% and after 700 °C annealing, which could be assigned to the formation of elemental Co nanocrystal clusters within the implanted layer [Norton 2003]. Such elemental nanoclusters could be below the sensitivity of the Raman and XRD experiments of this thesis.

For Co- and Mn-alloyed ZnO, polycrystalline bulk samples were available grown by a VPT technique. Their concentration is found to be hard to control and inhomogeneous, and their crystal orientation is undefined. While no secondary phases could be identified for these samples with concentrations ≤5 at.% by XRD and by room temperature Raman scattering, magnons of CoO and electronic Raman scattering due to Co^{2+} ions in CoO were observed in low temperature Raman experiments of the Co-alloyed bulk samples. The reason of the TM oxide formation at such low concentration can be attributed to the VPT growth technique.

ZnO:Mn nanoparticles and ZnO:Co nanocrystalline layers were fabricated to study the effect of

secondary phase	magnetic ordering	reference
Mn (clusters)	Paramagnet, (Superpara-)	[Knickelbein 2001]
MnO	Antiferromagnet	[Ip 2003]
MnO_2	Antiferromagnet	[Ip 2003]
Mn_3O_4	Ferrimagnet	[Zheng 2004]
$ZnMn_2O_4$	Ferrimagnet	[Zheng 2004]
Co (clusters)	Ferromagnet, (Superpara-)	[Billas 1994, Park 2004]
CoO	Antiferromagnet	[Zhou 2007/1]
Co_3O_4	Antiferromagnet	[Zhou 2007/1]
$ZnCo_2O_4$	Paramagnet	[Kim 2004]
Fe (clusters)	Ferromagnet, (Superpara-)	[Billas 1994]
Fe_3O_4	Ferrimagnet	[Zhou 2007/2]
$ZnFe_2O_4$ (nanocry.)	Ferrimagnet	[Zhou 2007/2]
$ZnFe_2O_4$ (bulk)	Antiferromagnet	[Zhou 2007/2]
Ni (clusters)	Ferromagnet, (Superpara-)	[Billas 1994, Slater 1936]
NiO	Antiferromagnet	[Kodama 1997, Roth 1958]

Table 5.8: *Magnetic properties of identified or potential secondary phases in TM-alloyed ZnO.*

TM alloying on low-dimensional ZnO systems. It was found that in a wet-chemical synthesis the resulting position of added TM ions is hard to control and therefore TM clustering with the organic capping molecules becomes a risk. Even if the TM ions are incorporated in the nanocrystalline core, the results on ZnO:Co nanocrystalline layers show, that the self purification tendency of such nanosystems leads to secondary phase formation already at low TM concentrations. Thus, the room temperature ferromagnetism of these layers observed for higher concentrations is most likely due to elemental Co clusters.

Besides the identification of secondary phases, also indications of substitutional TM incorporation were found. In the case of Mn, for example, a potential candidate for an isolated impurity mode was identified and the EPR signature of substitutional Mn incorporation was observed. In addition, the luminescence of tetrahedrally coordinated Co^{2+} ions was observed in the red spectral range for all Co-alloyed samples.

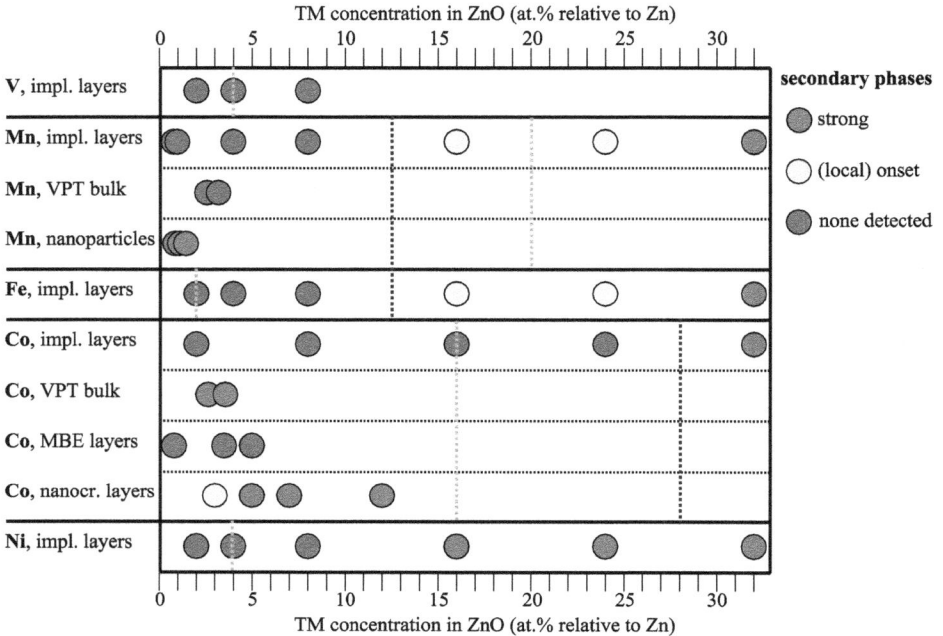

Figure 5.41: *Observed solubility limits of the TM-alloyed ZnO systems studied for this thesis. Green circles denote samples where no secondary phases were detected by Raman scattering and the applied complementary methods. Samples with only few localized secondary phase inclusions are labeled with yellow circles. Red circles, finally, are used for strong secondary phase formation. Additionally, reported solubility limits for the TM species are drawn in the diagram with vertical lines: blue lines are used for the solubility limits reported for MBE samples [Jin 2001], orange lines correspond to the solubility limits of TM-alloyed ZnO fabricated by a solid-state reaction technique [Kolesnik 2004]. In the case of the TM-implanted samples and the nanocrystalline ZnCoO layers, this diagram shows the secondary phase properties after 700 °C annealing.*

In summary, the magnetic properties of the studied TM-alloyed ZnO systems are often dominated by secondary phases. In Figure 5.41, the solubility limits for the TM-alloyed samples in this thesis are displayed. Obviously, the onset of secondary phase formation is not only determined by the TM species and concentration, but also by the sample dimensionality and the fabrication process. While it is still plausible that there are substitutional TM ions on Zn sites in all samples, the origin of reported ferromagnetism at room temperature is ambiguous, especially for low-dimensional systems and systems with high TM concentrations >10 at.%. Such magnetic findings have to be supplemented with results on the actual TM ion position and on the presence of secondary phases, using methods with high sensitivity for very small inclusions, especially of elemental TM clusters and TM oxides. Recently, these findings were confirmed by similar studies at the Forschungszentrum Dresden-Rossendorf, where nanocrystalline secondary phase inclusions were identified in Mn-, Fe-, Co-, and Ni-alloyed ZnO and conventional XRD alone was found to be not sufficient for the identification of such precipitates [Zhou 2008/1, Zhou 2008/2, Zhou 2008/3].

Chapter 6

Nitrogen-doped ZnO

The technical issues of p-type doping of ZnO and its importance for (opto-)electronic applications were discussed in subsection 3.2.1. One promising acceptor candidate is substitutional nitrogen on oxygen sites (N_O), which was already successfully employed for the fabrication of ZnO-based blue LEDs [Tsukazaki 2004]. For studying the structural properties of doped ZnO, Raman spectroscopy is one of the most commonly used characterization methods. Several additional Raman modes were observed in the Raman spectra of N-doped ZnO [Wang 2001]. Subsequently, various other Raman studies on ZnO with incorporated nitrogen were published. However, the origin of these modes is still ambiguous and strongly disputed.

For the studies in this thesis, nitrogen incorporation into ZnO was achieved by ion implantation (section 6.1) and during MBE growth (section 6.2). The main question is whether the additional vibration modes are local modes of substitutional nitrogen on oxygen sites or if they correspond to disorder-induced Raman scattering. For this reason, a systematic Raman investigation was performed, whose results will be discussed in detail in this chapter. The additional modes as well as the structural impact of nitrogen incorporation on the ZnO crystal are studied and the samples fabricated by implantation and by epitaxial growth are compared. All Raman scattering experiments discussed in this chapter were conducted in micro-Raman configuration at room temperature with the 514.5 nm line of an argon ion laser. The used scattering configuration was 180° backscattering with incident and scattered light directions along the ZnO c-axis where not stated otherwise.

6.1 Nitrogen doping of ZnO by ion implantation

6.1.1 Experimental results

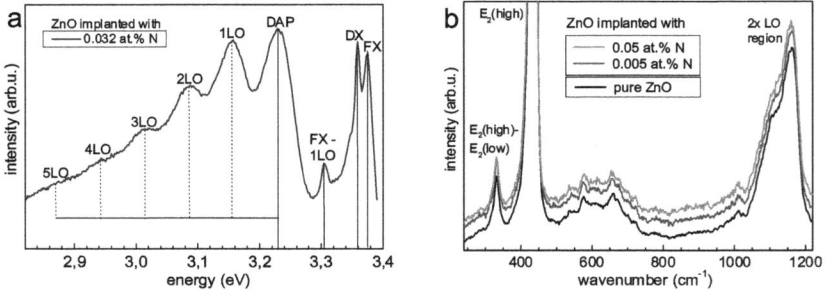

Figure 6.1: *(a) Photoluminescence spectrum of ZnO implanted with 0.032 at.% N, showing a strong donor-acceptor pair transition band at about 3.23 eV and the corresponding phonon replica (excitation: λ = 325.0 nm). (b) The Raman spectra of 0.005 at.% and 0.05 at.% nitrogen-implanted ZnO (after annealing) are nearly identical to the spectrum of pure ZnO (excitation: λ = 514.5 nm).*

Hydrothermally grown ZnO single crystals from CrysTec (Berlin) with (0001) orientation were implanted with nitrogen ions. The implanted N concentrations studied for this thesis are 0.005 at.%, 0.032 at.%, 0.05 at.%, 1 at.%, 2 at.%, and 4 at.% relative to O within a box-like implantation profile of about 200-300 nm. 0.005% relative to O, for example, corresponds to a volume density of about 2×10^{18} cm^{-3}. After the implantation process, all samples were annealed at 600 °C in vacuum for 30 min. For the 1 at.%, 2 at.%, and 4 at.% samples, an additional 800 °C annealing was applied.

The acceptor effect of incorporated nitrogen in ZnO crystals is reflected in their PL spectra by the occurrence of a donor-acceptor pair (DAP) band [Meyer 2004, Reuss 2004, Tamura 2003]. This band corresponds to electron-hole recombinations in donor-acceptor pairs after optical excitation of a semiconductor above its band gap [Thomas 1964]. As described in subsection 3.2.1, ZnO crystals are intrinsic n-type and the presence of abundant donor levels can be expected. Therefore, the intensity of the DAP transitions reflects the number of acceptor levels created by the nitrogen incorporation. In the PL spectra of as-grown ZnO, no DAP transitions are observed. In contrast, the DAP band at about 3.23 eV and its phonon replica occur very strong in the low temperature PL spectrum of the 0.032 at.% N-implanted ZnO in Figure 6.1a. This spectrum was recorded after 600 °C vacuum annealing, which proved to be preferable to annealing in air [Dürr 2008]. The PL experiment was conducted at 15 K with the 325.0 nm line of a He-Cd laser, which implies strong absorbance in the ZnO crystal and therefore high surface sensitivity. While the strong DAP band intensity confirms the presence of N_O acceptors, the systems could not be identified as p-type because the required transport experiments were complicated by the system geometry. The properties of the implanted layer of sub-μm depth strongly depend on the n-type ZnO host crystal. Nevertheless, p-type behavior was reported for N-implanted ZnO [Lin 2004, Tsai 2008] and, in any case, these nitrogen-implanted layers are well-suited model systems to study the optical and structural properties of N-doped ZnO.

In Figure 6.1b, the three Raman spectra of pure ZnO, 0.005% N-, and 0.05% N-implanted ZnO show the same features. They can all be assigned to the characteristic phonon modes of pure ZnO presented in subsection 3.1.2: E_2(high)-E_2(low) at about 332 cm^{-1}, E_2(high) at about 438 cm^{-1}, very weak A_1(LO) at about 576 cm^{-1}, and second order features between 1000 cm^{-1} and 1200 cm^{-1}. Other weak peaks correspond to multi-phonon modes as presented in Table 3.1 in subsection 3.1.2. No additional modes are observed at such low nitrogen concentrations. This seems to contradict the findings of Kaschner et al., who observed strong additional Raman modes for N concentrations of only 0.0025 at.% [Kaschner 2002]. However, the wavelength of 514.5 nm used for the Raman scattering experiments implies that the Raman signal is derived not solely from the implanted layer, but mostly from the underlying intact ZnO crystal. In contrast, the samples used in the work of Kaschner et al. were grown by chemical vapor deposition and nitrogen is incorporated throughout the ZnO crystal.

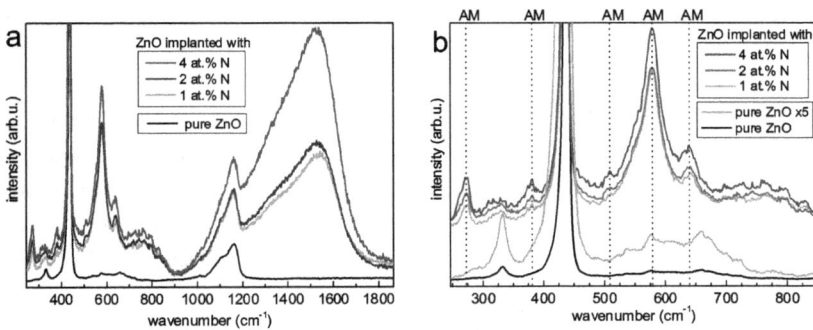

Figure 6.2: (a) Raman spectra of 1 at.%, 2 at.%, and 4 at.% nitrogen-implanted ZnO before annealing, compared to pure ZnO (excitation: $\lambda = 514.5$ nm). Upon N implantation, strong additional modes and a broad background signal occur between 250 cm^{-1} and 900 cm^{-1} and a very broad band evolves between 1000 cm^{-1} and 1800 cm^{-1}. (b) Raman spectra of (a) shown in the lower-frequency region. Additional modes (AM) are marked by vertical lines.

To study the effect of N incorporation on the Raman spectra of the implanted samples, higher concentrations are required. For this purpose, the 1 at.%, 2 at.%, and 4 at.% N-implanted ZnO crystals were fabricated. Not only the expected Raman intensity of potential additional modes is higher due to the increased nitrogen concentration, but also the relative contribution of the implanted layer to the total Raman signal increases. By comparison with literature data on ZnO implanted with comparable N concentrations [Friedrich 2007, Reuss 2004], the sensitivity of the Raman setup used for this thesis is found to be very competitive. Note that these nitrogen concentrations are still in an applicable range. For example, Tsukazaki et al. achieved p-type ZnO with carrier concentration in the 10^{16} cm^{-3} range by a nitrogen concentration of about 0.5 at.% [Tsukazaki 2004]. Figures 6.2a and 6.2b show the 1 at.%, 2 at.%, and 4 at.% N-doped ZnO crystals after implantation and without thermal annealing compared to pure ZnO. Two frequency regions are especially affected by the nitrogen implantation: Strong additional modes and a broad background signal occur between 250 cm^{-1} and 900 cm^{-1} and a very broad band evolves from 1000 cm^{-1} to 1800 cm^{-1}, superimposing the ZnO two-phonon features at 1000-1200 cm^{-1}. Besides the broad signals, sharp additional modes can be identified at about 273 cm^{-1}, 381 cm^{-1}, 509 cm^{-1}, 579 cm^{-1}, and 640 cm^{-1}. Since there are no modes with 273 cm^{-1}, 509 cm^{-1}, and 640 cm^{-1} in pure ZnO or in ZnO implanted with other ions, these Raman features can be identified as nitrogen-related. The modes at about 380 cm^{-1} and 580 cm^{-1} are located near the frequency positions of the A_1(TO) mode and the A_1(LO) mode of pure ZnO, respectively. In implanted and unannealed ZnO, the crystal disorder can lead to reduced symmetry. Therefore, modes can be observed, which are symmetry-forbidden in the used scattering configuration. This is possibly the case for the feature at 380 cm^{-1}, which could correspond to the A_1(TO) mode. Such modes would be expected to disappear after thermal annealing in contrast to modes related to the incorporated impurities.

In chapter 5, resonance effects due to disorder-enhanced Fröhlich scattering were observed, especially for LO modes. Therefore, the strong additional mode near the A_1(LO) frequency can hardly be distinguished from the broad disorder band, which was identified for all implanted and disordered samples in chapters 4 and 5. Again, such a disorder signal is expected to be strongly reduced by thermal treatment and, therefore, an unambiguous assignment should be able upon annealing.

Figures 6.3a and 6.3b show the Raman spectra of the 1 at.%, 2 at.%, and 4 at.% N-implanted samples after 600 °C thermal annealing in vacuum for 30 min. Vacuum annealing at this temperature was found to be ideal for N activation, as deduced from the DAP signature in PL experiments [Dürr 2008]. The feature at about 380 cm^{-1}, observed before annealing in Figure 6.2, has disappeared and can therefore be attributed to the A_1(TO) mode, which obeys the symmetry selection rules in this more ordered system after annealing. Moreover, the background in

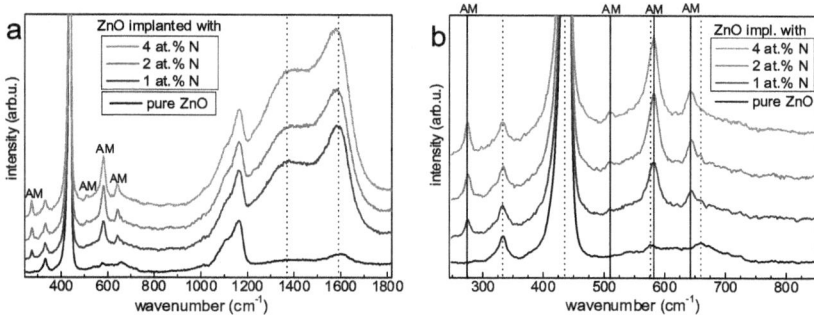

Figure 6.3: *(a) Raman spectra of 1 at.%, 2 at.%, and 4 at.% nitrogen-implanted ZnO after 30 min annealing at 600 °C in vacuum, compared to pure ZnO (excitation: $\lambda = 514.5$ nm). Again, the additional modes between 250 cm^{-1} and 700 cm^{-1} and the broad band between 1000 cm^{-1} and 1800 cm^{-1} occur. The broad band is also indicated in the spectrum of the pure ZnO sample. (b) Raman spectra of (a) shown in the lower-frequency region. Additional modes are marked by solid vertical lines, intrinsic ZnO modes are marked by dotted vertical lines.*

the 500-900 cm^{-1} region is strongly reduced as it can be expected for an improved crystalline quality. In contrast, the additional modes at 274 cm^{-1}, 510 cm^{-1}, 582 cm^{-1}, and 643 cm^{-1} are still clearly observed. The 582 cm^{-1} mode near the $A_1(LO)$ position is narrower than the disorder band observed in other implanted systems in the chapters 4 and 5. Moreover it is observed at a slightly higher wavenumber position ($\approx +5$ cm^{-1}) than the weak $A_1(LO)$ mode in the spectrum of the pure ZnO. This indicates a different origin of the 582 cm^{-1} additional mode than the $A_1(LO)$ phonon, what will be confirmed in subsection 6.1.2. The higher-frequency shoulder of the 643 cm^{-1} additional mode corresponds to a multi-phonon mode of ZnO as indicated by the vertical line in Figure 6.3b.

The strong band between 1200 cm^{-1} and 1700 cm^{-1} is still observed after the 600 °C annealing for all concentrations. However, two shoulders are resolved now in contrast to the single broad band observed before annealing. Interestingly, this two-shoulder signature is also indicated in the spectrum of the pure ZnO crystal after 600 °C annealing.

The Raman spectra of the 1 at.%, 2 at.%, and 4 at.% samples after a second annealing step in vacuum are displayed in Figures 6.4a-c. With 800 °C annealing, the crystal disorder has been further decreased and substitutional incorporation might be supported. However, a slight decrease of the DAP intensity was observed in the PL after annealing with temperatures higher than 600 °C.

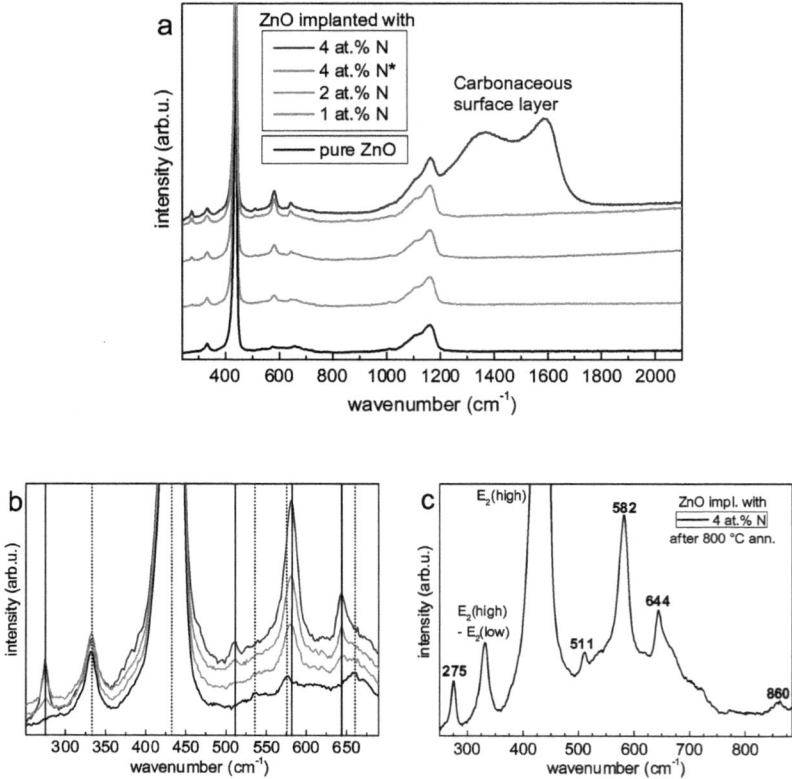

Figure 6.4: *(a) Raman spectra of 1 at.%, 2 at.%, and 4 at.% N-implanted ZnO after 30 min annealing at 800 °C in vacuum, compared to pure ZnO (excitation: $\lambda = 514.5\ nm$). The spectrum of the 4 at.% sample marked by an asterisk shows the Raman results after thorough surface cleaning using ethanol. The broad band between 1200 cm^{-1} and 1700 cm^{-1} has disappeared after this treatment and is identified as carbonaceous surface pollution. (b) Raman spectra of (a) shown in the lower-frequency region. Additional modes are marked by solid vertical lines, intrinsic ZnO modes are marked by dotted vertical lines. The intensity of the additional modes clearly scales with the nitrogen concentration. (c) Raman spectrum for the 4 at.% N-implanted ZnO crystal after 800 °C and with longer integration time (excitation: $\lambda = 514.5\ nm$). A fifth additional mode is resolved at about 860 cm^{-1}. The additional modes are labeled by their vibrational frequencies.*

First, the broad structure between 1200 cm^{-1} and 1700 cm^{-1} is discussed. Such a structure was not observed or reported for pure ZnO or nitrogen-doped ZnO. It resembles, however, the well-known Raman signature of carbonaceous material [Jawhari 1995]. Small carbon impurities were detected by SIMS measurements, but far below the detection limit of Raman scattering. This holds true even for carbonaceous material, which can show strong Raman resonance if in graphite or graphene structure (see Figure 2.4 in subsection 2.1.2). The spectrum of the 4 at.% N-implanted ZnO is shown twice in Figure 6.4a: before and after surface treatment with ethanol. Obviously, the broad signal between 1200 cm^{-1} and 1700 cm^{-1} has disappeared after this cleaning procedure. Therefore, this signal presumably corresponds to a surface pollution with carbonaceous material. This pollution may be caused by residua of the carbon pads used to fix the samples during implantation. While the pads were attached only to the sample backside, the pollution of the surface could have happened during the removal of the pads in alcoholic solution. Interestingly, this additional band was also observed in Figure 6.3a for the pure ZnO sample. This could be caused by contact of the pure ZnO sample with the implanted samples during sample processing (annealing, storage, transport).

The intensity of the additional modes clearly scales with the N concentration after the 800 °C annealing (Figure 6.4b). The additional mode at about 582 cm^{-1}, near the A$_1$(LO) mode position, is not significantly reduced despite the thorough annealing. This is in contrast to the behavior of the disorder band in this region observed in the implantation experiments in the chapters 4 and 5. Relative to the E$_2$(high) modes, the additional modes show slightly weaker intensity after each annealing step. However, as discussed above, annealing also improves the transparency of the implanted layers, and therefore, more Raman scattering signal derives from the substrate below. If this effect is not taken into account, the intensity reduction of the additional modes with annealing in N-implanted ZnO can result in misleading conclusions, e.g. in [Artus 2007]. In Figure 6.4c, a Raman spectrum of the 4 at.% N-implanted sample is shown, which was recorded with longer integration time. Besides the already mentioned additional modes at about 275 cm^{-1}, 511 cm^{-1}, 582 cm^{-1}, and 644 cm^{-1}, a weak feature occurs at about 860 cm^{-1}. The origin of these five additional modes is discussed in detail in subsection 6.1.2.

Reference	Additional Raman features (cm^{-1})					p-type
	AM1	AM2	AM3	AM4	AM5	
[Wang 2001]	274	508	581	642	857	no
[Kaschner 2002]	275	510	582	643	856	no
[Reuss 2004]	275	508	579	642	-	no
[Du 2005]	274	508	580	642	857	yes
[Haboeck 2005]	275	510	582	643	856	no
[Lu 2006]	275	506	579	642	-	no
[Tu 2006]	275	508	581	640	854	yes
[Wang 2006/2]	275	504	575	644	-	no
[Yu 2006]	275	510	576	640	-	no
[Artus 2007]	275	510	580	640	850	no
[Friedrich 2007]	275	510	577	-	-	no

Table 6.1: *Additional modes reported in various Raman studies on N-doped ZnO. The origin of the additional modes proposed by the authors is indicated by the text color: red = localized vibration of substitutional nitrogen; blue = intrinsic ZnO modes or nitrogen-related complexes.*

6.1.2   Discussion: Origin of the additional Raman features

The sharp additional modes in N-doped ZnO were reported for the first time in [Wang 2001] and since then observed for various N-doped systems during the recent years. Table 6.1 summarizes, in chronological order, the reported frequency values of the additional modes as well as the assignments made by the authors of the most prominent and most recent publications on this topic. As can be seen, the initial opinion was that the additional modes correspond to localized vibrational modes of substitutional nitrogen on oxygen sites. In recent publications, this opinion is not shared. Intrinsic ZnO modes or nitrogen-related complexes are favored for the explanation of these features.

Localized impurity modes due to substitutional nitrogen

After the first report in [Wang 2001], Kaschner et al. provided a first explanation of the additional modes in N-doped ZnO [Kaschner 2002]. They ascribed all five additional features to localized vibrations of nitrogen atoms on oxygen sites and based their suggestion mainly on two findings: (i) A successful nitrogen incorporation was confirmed by SIMS, and (ii) the intensity of the additional modes was found to scale with the nitrogen concentration. However, both findings would also be consistent with other explanations, for example with nitrogen complexes or ZnO defects induced by increasing nitrogen impurity concentration.

In subsection 2.3.2, theoretical considerations were presented to the question in which frequency regions localized vibrations of impurities can exist without coupling to eigenmodes of the surrounding crystal lattice. For ZnO, this is possible in the frequency gap between the acoustic and the optical phonon modes, from about 270 cm^{-1} to 410 cm^{-1}, and above the highest optical phonon mode frequency, i.e. above about 600 cm^{-1}. The frequency positions of the additional modes observed in subsection 6.1.1 are 275 cm^{-1}, 511 cm^{-1}, 582 cm^{-1}, 644 cm^{-1}, and 860 cm^{-1}. Therefore, the feature at about 275 cm^{-1} could be a gap mode and the features at about 644 cm^{-1} and 860 cm^{-1} could be local vibrational modes. In contrast, the 511 cm^{-1} mode is in a frequency region with high phonon density of states and the 582 cm^{-1} frequency position is near the LO phonon modes (see subsection 3.1.2). The mass of the substituting nitrogen atom is about 85% of the oxygen mass. As a consequence, generally higher vibrational frequencies can be expected compared to the existing ZnO modes. This contradicts the assignment of the 275 cm^{-1} feature to a gap mode of substitutional nitrogen.

In the related host material ZnSe, an experimental proof for localized vibrational modes of nitrogen was presented by the use of infrared absorption [Stein 1994]. ZnSe samples were implanted with the nitrogen isotope ^{14}N and for comparison alternatively with ^{15}N. As a result of the difference in mass, two vibrational frequencies were observed, at about 537 cm^{-1} and about 553 cm^{-1}. This result is in perfect agreement with the frequency ratio estimated in a diatomic model for localized vibrations of the N isotopes against Zn neighbors, in which case the relevant parameter is the reduced mass of both constituents [McCluskey 2000]:

$$\frac{\nu_{LVM}(^{15}N)}{\nu_{LVM}(^{14}N)} \simeq \sqrt{\frac{\mu(^{14}N, Zn)}{\mu(^{15}N, Zn)}} \simeq 0.97 \simeq \frac{537\ cm^{-1}}{553\ cm^{-1}}. \tag{6.1}$$

Analogously, ZnO crystals were implanted with ^{14}N and ^{15}N ions by Artus et al. and the frequency positions of the additional Raman features were analyzed [Artus 2007]. With the factor of 0.97 derived above, the frequency positions of the two strongest additional modes at about 275 cm^{-1} and 582 cm^{-1} could be expected to show a red-shift of about 8 cm^{-1} and 17 cm^{-1}, respectively, for the sample implanted with the ^{15}N isotope. Such shifts should be easily detectable with the typical experimental resolution below 2 cm^{-1} in conventional Raman setups. However, no frequency shifts occurred for these two strong modes. The same behavior was observed for the additional mode at about 510 cm^{-1}. Therefore it is excluded that the modes at 275 cm^{-1}, 510 cm^{-1}, and 582 cm^{-1} are localized impurity modes of substitutional nitrogen. The modes at 644 cm^{-1} and 860 cm^{-1} could not be observed in this work due to insufficient sensitivity. Furthermore, as stated above, they lie in a frequency region which is expected for local impurity modes of nitrogen. On the other hand, all five additional modes show the same intensity behavior upon annealing and for increasing nitrogen concentrations. Therefore, the same origin for all additional modes seems most probable.

Another possibility of localized vibration modes was suggested in [Wang 2006/2]. They calculated the local phonon density of states (LPDOS) of various possible configurations in N-substituted ZnO using a real-space approach. A vibrational mode with 275 cm^{-1} frequency was found in these calculations for a Zn atom if one of its nearest oxygen neighbors is substituted by nitrogen atoms. This vibration would not contradict the findings in the isotope experiments of Artus et al. because the vibration is localized at a Zn atom. However, this theoretical finding regarding the 275 cm^{-1} mode origin has not been supported experimentally up to now and only covers this single mode.

Nitrogen-related complexes

Since the possibility of localized vibrational modes of substitutional nitrogen on an O site was ruled out by isotopic studies, an explanation often used in literature is the formation of not further specified nitrogen-related 'complexes' [Artus 2007, Friedrich 2007]. In the following, several possible complexes are analyzed with regard to the Raman scattering results obtained in this thesis.

First, nitrogen-hydrogen complexes are discussed. In N-doped ZnSe, the N-H stretching and wagging vibrations of nitrogen-hydrogen complexes were observed by Raman scattering at about 3200 cm^{-1} and 800 cm^{-1}, respectively [Wolk 1993]. Haboeck et al. observed several high-frequency Raman features in N-doped ZnO between about 2200 cm^{-1} and 2300 cm^{-1}, which they assigned to the vibrations of lattice-bound NH_3 [Haboeck 2005]. The N-H stretching mode in ZnO:N was reported by Nickel et al. at about 3100 cm^{-1}, while other modes between 2800 cm^{-1} and 3000 cm^{-1} were assigned to C-H vibrations [Nickel 2003]. In contrast, no additional modes have been observed in the Raman spectra of N-doped ZnO in this thesis in the frequency range between 2000 cm^{-1} and 3200 cm^{-1}. The broad band between 1200 cm^{-1} and 1700 cm^{-1} in several spectra of subsection 6.1.1 could be ascribed to carbonaceous surface pollution. In addition, even 30 min annealing at 800 °C did not result in a disappearance of the additional modes, not even a significant intensity decrease could be detected. These findings contradict the involvement of hydrogen, which is expected to be volatile at such high temperatures and which leads to typical vibrational frequencies much higher than the observed additional modes. The formation of N_2 molecules on oxygen sites is not excluded in N-doped ZnO crystals [Lee 2001]. However, the calculated vibrational frequencies of such substitutional N_2 are about 1500 cm^{-1} and 2150 cm^{-1} [Limpijumnong 2005]. In the same publication, also the vibrations of NO_O are simulated (950 cm^{-1} and 1500 cm^{-1}).

In summary, the expected strongest modes for the most probable nitrogen and nitrogen-hydrogen complexes in ZnO are located at high wavenumber values due to the low masses of N and H. In these high frequency regions, no indications of additional modes are observed in the Raman spectra for this thesis. Thus, it is not likely that the observed additional modes at low wavenumber values are caused by such complexes.

Disorder-induced Raman scattering

The most widely recognized publication on the origin of the additional Raman modes in N-doped ZnO is a work by Bundesmann et al. They suggest that the additional modes correspond to disorder-induced Raman scattering, independent from the species of the disorder-producing impurity [Bundesmann 2003]. Experimental evidence for this suggestion was presented in form of Raman scattering results on Fe-, Sb-, Al-, Ga-, and Li-alloyed ZnO. Features near or exactly at the frequency positions of the additional modes were most strongly seen in the Raman spectrum of Al-alloyed ZnO and in the spectrum of one of the two studied Ga-alloyed samples. In addition, smaller features near the discussed wavenumber values were also observed for the Fe- and the Sb-alloyed samples. The Li-alloyed sample showed no additional modes.

These findings are in marked contrast to the findings of this thesis: Among all ZnO-based systems with various impurity elements, concentrations, fabrication techniques, and after-growth treatments, these modes are only seen in nitrogen-doped ZnO systems. Especially, ZnO implanted with Ar, Mn, Ni, V, Co, or Fe did not reveal any indication of these modes. These findings are confirmed by numerous Raman studies of other research groups. For example, additional modes were observed in N-implanted ZnO, but not in samples implanted with P, O, Zn [Artus 2007], O, Si, Ga [Yu 2006], Ga [Reuss 2004], and H [Friedrich 2007], respectively. The implantation and annealing studies in subsection 4.1.1 and section 5.1 demonstrate that a broad and intense band in the LO region is characteristic for disorder-induced Raman scattering in ZnO, reflecting the phonon density of states and disorder-enhanced Fröhlich scattering of the $A_1(LO)$ mode. In these experiments, no additional modes are induced by the structural disorder.

How could the findings of Bundesmann et al. be explained in this context? The strongest additional modes were observed for Al alloying and in one of the ZnO:Ga samples, while the other Ga-alloyed sample shows much weaker additional mode signature. The difference between the two Ga-alloyed ZnO crystals was the atmosphere during growth. The sample with the weak additional modes was grown in oxygen atmosphere, the sample with the strong additional modes in nitrogen atmosphere. This suggests that nitrogen was accidentally incorporated during growth and induced the strong modes in the respective Ga-alloyed sample. In addition, Bundesmann et al. can not exclude nitrogen concentrations below 3 at.%. As can be seen from the results in subsection 6.1.1, such concentrations could easily account for the observed additional modes. The weak additional modes in the Fe-, Sb-, and Ga-alloyed sample are therefore attributed to unintentional nitrogen incorporation during growth.

An interesting case is the Al-alloyed sample. It was grown in oxygen atmosphere, but still shows very strong additional modes at about 276 cm^{-1}, 509 cm^{-1}, 579 cm^{-1}, and 643 cm^{-1}. In contrast to Fe-, Sb-, and Ga-alloyed ZnO, this finding for Al-alloyed ZnO is confirmed by the experimental studies in [Choopun 2007, Tzolov 2000, Tzolov 2001]. However, no additional modes are observed in other Raman studies on Al-alloyed ZnO systems, for example in [Behera 2008].

To summarize the experimental situation, the additional modes are reported in most nitrogen-alloyed ZnO systems, in few studies on Al-alloyed ZnO, and for several alloy elements in the work of Bundesmann. They are not observed in numerous other studies on disordered ZnO or ZnO alloyed with other impurities than N or Al. The Raman signature of the nitrogen-doped systems does not agree with the Raman spectra generally seen in disordered ZnO.

Impurity-activated silent modes

Using an ab initio method, Manjon et al. calculated the vibrational frequencies of the B_1(low) and B_1(high) mode of ZnO [Manjon 2005]. These modes are silent modes, i.e. not observable in Raman scattering and infrared absorption under normal conditions. The computed vibrational frequencies of B_1(low) and B_1(high) are 265 cm^{-1} and 552 cm^{-1}, respectively. While the frequency of the B_1(low) mode agrees quite well with the additional mode observed at about 275 cm^{-1}, the frequency difference between the B_1(high) calculation and the additional mode at about 582 cm^{-1} is rather high. Nevertheless, several arguments support such an assignment. The applied ab initio method calculates the A_1(LO) and E_1(LO) phonon modes of ZnO with about 550 cm^{-1} and 560 cm^{-1}, respectively [Serrano 2004], compared to their experimentally observed frequencies of about 574 cm^{-1} and 590 cm^{-1}, respectively [Cusco 2007]. This underestimation of about 24 cm^{-1} to 30 cm^{-1} could explain the frequency discrepancy between the calculated B_1(high) mode at about 552 cm^{-1} and the observed additional mode at about 582 cm^{-1}. Provided that this silent mode assignment is correct for the 275 cm^{-1} and 582 cm^{-1} additional modes, the other modes could be explained by multi-phonon processes involving the silent modes. Taking the experimental data, B_1(high)+B_1(low) could be expected at about 857 cm^{-1} with much weaker intensity, in good agreement with the additional mode at about 860 cm^{-1}. The modes at about 511 cm^{-1} and 644 cm^{-1} are assigned to 2xB_1(low) and B_1(high)+TA. The experimental data presented in subsection 6.1.1 support this assignment. The two modes assigned to B_1(low) and B_1(high) show the strongest intensity. Furthermore, these modes are narrow compared to the disorder-induced Raman scattering observed in many TM-alloyed samples in chapter 5. The additional modes persist even after thorough annealing, which contradicts the hypothesis of pure, structural disorder as origin.

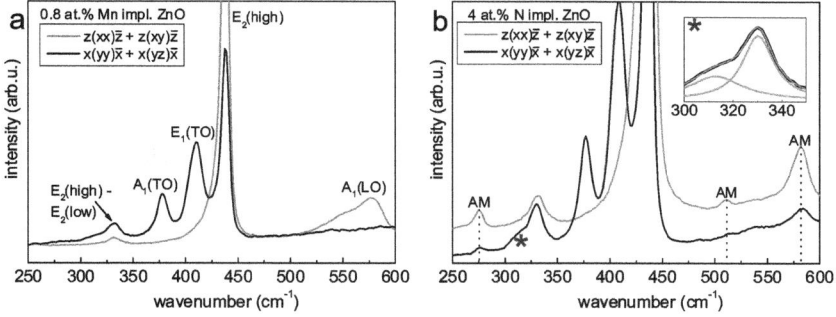

Figure 6.5: *(a) Raman spectra of 0.8 at.% Mn-implanted ZnO without annealing, recorded in different scattering configurations (excitation: $\lambda = 514.5$ nm). The disorder-enhanced $A_1(LO)$ mode is symmetry-forbidden if the directions of incident and scattered light are perpendicular to the ZnO c-axis, i.e. $x(...)\bar{x}$. (b) Raman spectra of 4 at.% N-implanted ZnO after 800 °C annealing, recorded in different scattering configurations (excitation: $\lambda = 514.5$ nm). In both configurations, all additional modes are visible, including the one with vibrational frequency similar to the $A_1(LO)$ phonon mode.*

The assignment of the signal at about 582 cm^{-1} to the B$_1$(high) phonon mode is not shared by most publications, where it is assigned to the A$_1$(LO) mode. In fact, the A$_1$(LO) mode frequency is very near to this additional feature (about 574 cm^{-1}) and is known to be sensitive to disorder as shown in chapters 4 and 5. The strong intensity and the frequency position of this additional mode might suggest that it is the A$_1$(LO) mode of ZnO, disorder-enhanced by nitrogen incorporation. Friedrich et al., for example, compare nitrogen- and hydrogen-implanted ZnO without thermal annealing [Friedrich 2007]. In both samples, the disorder band in the A$_1$(LO) region is observed, in agreement with the implantation studies of this thesis (e.g. subsection 4.1.1). From this fact, they conclude that the additional mode reported for nitrogen in this region is identical with the A$_1$(LO) mode. As will be shown in the following, this assignment is quite arguable. The mode is not reduced to nearly zero despite a thorough annealing at 800 °C. Furthermore, it is narrower than the broad disorder band.

To clarify the origin of this additional mode at about 582 cm^{-1}, Raman experiments were conducted with different scattering configurations: (i) incident and scattered light directions along the ZnO c-axis, i.e. $z(...)\bar{z}$ and (ii) incident and scattered light directions perpendicular to the c-axis, i.e. $x(...)\bar{x}$. To be sensitive for diagonal as well as off-diagonal Raman tensor components, no polarization selection of the scattered light was used. The A$_1$(LO) mode is symmetry-forbidden in both $x(yy)\bar{x}$ and $x(yz)\bar{x}$ configuration [Cusco 2007]. Figure 6.5a shows the Raman spectra of 0.8 at.% Mn-implanted ZnO, recorded in $z(...)\bar{z}$ and $x(...)\bar{x}$ configuration, respectively. This sample was chosen because no secondary phases are expected or indicated and, in the usual scattering configuration, it shows the broad band in the $A_1(LO)$ region characteristic for disordered ZnO. As expected from the selection rules, this $A_1(LO)$ band is not observed in the $x(...)\bar{x}$ scattering configuration. Only the ZnO modes allowed for this symmetry are observed, i.e. the $E_2(high)$-$E_2(low)$ at about 330 cm^{-1}, the $A_1(TO)$ at about 377 cm^{-1}, the $E_1(TO)$ at about 410 cm^{-1}, and the $E_2(high)$ at about 437 cm^{-1}. These ZnO modes are also observed in the $x(...)\bar{x}$ configuration for the 4 at.% nitrogen-implanted sample in Figure 6.5b. But in addition, the modes at about 275 cm^{-1}, 510 cm^{-1}, and 580 cm^{-1} are visible, too. Therefore, it is clear that the additional mode at about 580 cm^{-1} has a different origin than the well-known disorder band. Presumably, it is the activated $B_1(high)$ and only coincides with the $A_1(LO)$ frequency position.

Besides the observed silent modes and silent-mode-related multi-phonon features, an additional feature is predicted in [Manjon 2005]. The $B_1(high)$-$B_1(low)$ feature can be expected at about 582 cm^{-1} minus 275 cm^{-1}, near the $E_2(high)$-$E_2(low)$ phonon mode. In fact, the difference mode can be observed as a lower-frequency shoulder of the $E_2(high)$-$E_2(low)$ mode in Figure 6.5b. A two-Lorentzian fit gives a frequency of about 310 cm^{-1} for this shoulder assigned to the B_1 difference mode. Note that this difference mode is allowed even in pure ZnO and was observed with very weak intensity in Raman scattering [Cusco 2007].

Further support for the mode assignment comes from the observation of disorder-activated scattering of B_1 silent modes in other wurtzite-type materials such as InN, GaN, AlN [Manjon 2005]. As to the reason for the B_1 silent mode activation in the case of nitrogen, Manjon et al. suggest the strong change of the electronic properties between nitrogen and the substituted oxygen due to central-cell corrections, which especially affect elements of the first row of the table of the elements. Following this reasoning, the absence of additional features in the TM-alloyed ZnO systems in this thesis could be explained because Zn is substituted by the electronically similar TM elements V, Mn, Fe, Co, and Ni. Tzolov et al. explained the occurrence of additional modes at about 274 cm^{-1} and 580 cm^{-1} in Al-alloyed ZnO with the carrier introduction by Al doping and attributed the Raman signals to silent mode activation by electric-field-induced Raman scattering [Tzolov 2000, Tzolov 2001].

Conclusion of the additional mode discussion

To conclude, additional modes were observed in the Raman spectra of nitrogen-doped ZnO at about 275 cm^{-1}, 511 cm^{-1}, 582 cm^{-1}, 644 cm^{-1}, and 860 cm^{-1}. They are assigned to the impurity-activated silent modes $B_1(low)$ and $B_1(high)$ and related multi-phonon modes. Since they are not observed in other ZnO systems in this thesis, the silent mode activation can be described as characteristic for nitrogen, but it is not excluded for other elements. Regarding literature results, another candidate could be aluminum-doped ZnO, for example. The mechanism behind the silent mode activation by nitrogen incorporation stays unclear. Possible directions could be (i) the different electronic properties of nitrogen and the substituted oxygen or (ii) a correlation with the doping effect of nitrogen on oxygen sites. If the mechanism is clarified in future work, the additional modes might be used as confirmation of substitutional nitrogen on oxygen sites. Further insight in the connection between the additional modes and other structural and optical properties of N-doped ZnO was achieved in the experiments with MBE-grown samples presented in section 6.2.

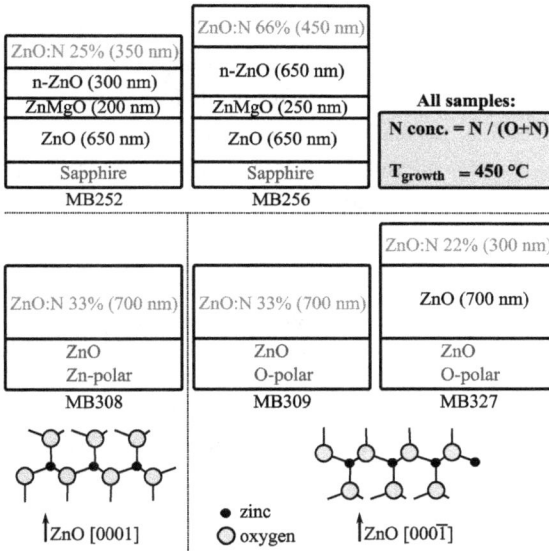

Figure 6.6: *Overview over the MBE-grown, nitrogen-doped ZnO samples. MB252 and MB256 were heteroepitaxially grown on sapphire substrate. The not shown MB254 differs from MB256 only by the nitrogen concentration (40% and 66%, respectively). MB308 and MB309 were grown homoepitaxially on Zn-polar and O-polar ZnO substrate, respectively. MB327 consists of a nitrogen-doped ZnO layer on top of a high-quality ZnO layer, grown on O-polar ZnO substrate.*

6.2   Nitrogen doping of ZnO by epitaxial growth

Nitrogen incorporation by ion implantation, as discussed in section 6.1, induces disorder, which is known to add to the intrinsic n-type properties of ZnO. Epitaxial growth by MBE, on the other hand, permits nitrogen incorporation into ZnO without irradiation damage.

In a heteroepitaxial growth series, N-doped ZnO samples were grown on sapphire substrate with 'nitrogen concentrations' n_N of 25%, 40%, and 66%, referred to as MB252, MB254, and MB256, respectively. Additionally, the homoepitaxial samples MB308, MB309, and MB327 were grown with 'nitrogen concentrations' n_N from 22% to 33%. For both series, n_N denotes the fraction of nitrogen molecules to the total number of nitrogen and oxygen molecules in the gas phase: n_N = N_2 / (N_2 + O_2). Figure 6.6 shows the designs of these samples. The not shown MB254 differs from MB256 only by the nitrogen concentration. All samples were grown at 450 °C. Further details concerning the experimental setup are provided in [Bakin 2006].

This section again focuses on the additional Raman features of nitrogen-doped ZnO. Furthermore, the influence of the different growth conditions on the structural properties of the ZnO:N top layers and on the doping effect of nitrogen is studied.

6.2.1   Nitrogen-doped ZnO, grown by heteroepitaxy

Before the spectroscopy experiments, the surface topology of the heteroepitaxially grown samples MB252, MB254, and MB256 were studied by optical microscopy (50x). The surface of MB252, grown with 25% nitrogen in the gas phase, is mostly homogeneous with only few dark spots of μm size. This changes with increasing N concentration in the gas phase. The density of the black spots increases from MB252 to MB254 (40%), and from MB254 to MB256 (66%). Surface pictures of MB252 and MB256 are shown in Figure 6.7, together with micro-Raman scattering results recorded while focusing on different spots of the sample surface. Besides the well-known ZnO phonon modes, the Raman spectra of the dark spots exhibit strong additional modes at about 274 cm^{-1}, 510 cm^{-1}, 581 cm^{-1}, 643 cm^{-1}, and 856 cm^{-1}, which were identified as activated B_1 silent modes and their corresponding multi-phonon processes in section 6.1. The inhomogeneous broadening of the mode at about 581 cm^{-1} and the strong background between 400 cm^{-1} and 700 cm^{-1}, especially in MB256, strengthen the impression of a poor structural quality within these spots. The spectra taken with focus on the surrounding surface show no additional features, but only ZnO phonon modes. Nevertheless, the occurrence of the A_1(TO) mode at about 378 cm^{-1} and the rather strong A_1(LO) mode at about 577 cm^{-1} indicate reduced structural quality in these regions as well. In the used 180° backscattering

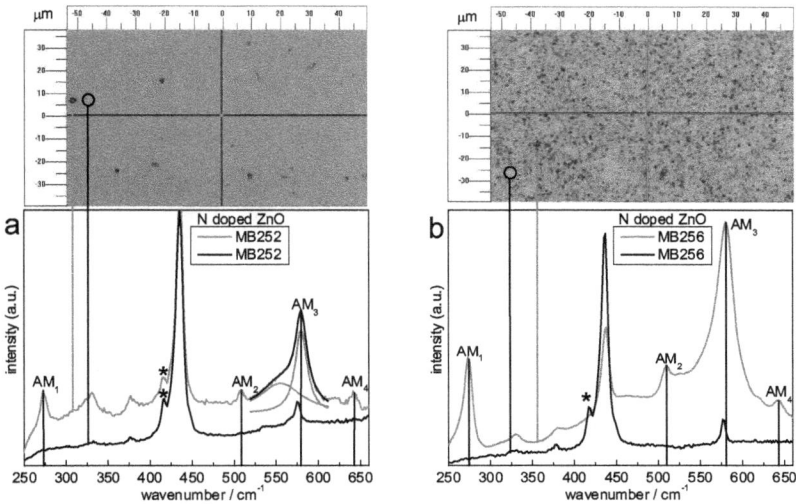

Figure 6.7: *Raman spectra of (a) MB252, grown with 25% nitrogen in the gas phase, and (b) MB256, grown with 66% nitrogen in the gas phase (excitation: λ = 514.5 nm). The microscope pictures show the surface spots where the laser was focused during the Raman scattering experiments. The additional modes occur in the spectra recorded with focus on the dark islands and are related to activated B_1 silent modes of ZnO. Signals marked by an asterisk come from the sapphire substrate.*

configuration with incident and scattered light directions along the ZnO c-axis, the $A_1(TO)$ is symmetry-forbidden and the $A_1(LO)$ is usually observed with very weak intensity in well-ordered ZnO [Cusco 2007]. Note that the $A_1(TO)$ occurs even stronger in the spectra taken with focus on the dark islands. These findings show that, under the used growth conditions, nitrogen incorporation is inhomogeneous and induces significant disorder in the ZnO:N top layer. The disorder clearly scales with the used nitrogen concentration in the gas phase. The additional modes are only seen when focused on the dark spots, indicating a stronger nitrogen incorporation than in the surrounding surface regions. To further study this aspect and in order to fabricate ZnO:N samples with improved structural quality, homoepitaxial samples were grown, which will be discussed in subsection 6.2.2.

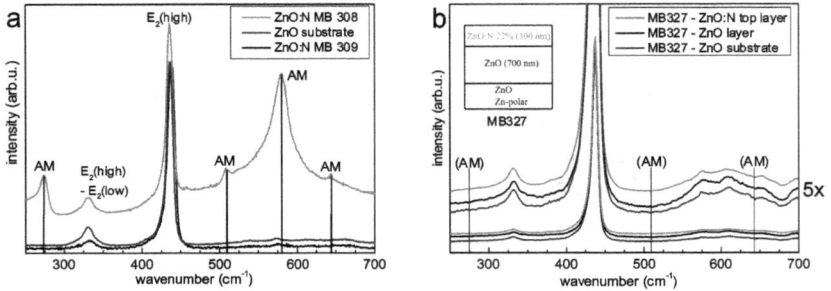

Figure 6.8: *(a) Raman spectra of MB308 and MB309, compared to pure ZnO substrate (excitation: $\lambda = 514.5$ nm). Both samples were grown with 33% N in the gas phase, MB308 on Zn-polar ZnO substrate, MB309 on O-polar substrate. For MB308, additional modes are observed, indicating nitrogen incorporation. (b) Raman scattering on the different layers of MB327 (excitation: $\lambda = 514.5$ nm). The spectra of the O-polar ZnO substrate, the high-quality ZnO buffer layer, and the ZnO:N top layer are identical within the experimental sensitivity. In particular, no additional modes are observed in the top layer, despite 22% nitrogen in the gas phase.*

6.2.2   Nitrogen-doped ZnO, grown by homoepitaxy

For the samples MB308 and MB309, a ZnO:N top layer was grown directly on Zn-polar and O-polar ZnO, respectively. For both samples, a nitrogen concentration in the gas phase of 33% was chosen. In the microscope pictures (50x), the surfaces of both samples were found to be homogeneous. In particular, no dark islands were observed as seen on the surface of the heteroepitaxial samples in subsection 6.2.1. Nevertheless, huge differences occur in the Raman spectra of the samples, as shown in Figure 6.8a. While the spectrum of MB309 only reveals the Raman features of pure ZnO, the additional, B_1-related modes are observed at about 274 cm^{-1}, 510 cm^{-1}, 580 cm^{-1}, and 644 cm^{-1} for MB308. In both samples, no symmetry-forbidden modes occur.

To evaluate these findings, the results of transport experiments are taken into account. The top layer of MB309 shows n-type conductivity like undoped ZnO. In contrast, the ZnO:N layer on top of MB308 is semi-insulating, indicating that the nitrogen incorporation led to a compensation of the ZnO-intrinsic donor levels. This partly successful doping is accompanied by the occurrence of the additional Raman features. This may indicate that the additional Raman modes could be suited to test substitutional nitrogen incorporation after all. However, the mechanism behind the activation of the silent B_1 modes should be understood in advance.

In addition, it can be stated that growth of ZnO:N layers on Zn-polar substrate is the more promising growth procedure. In this context, Lautenschlaeger et al. recently reported that the structural quality of ZnO epilayers grown on Zn-polar and O-polar ZnO substrate show equivalent structural properties within the experimental sensitivity of AFM and XRD, but differ in the optical properties [Lautenschlaeger 2008]. Additional excitonic recombinations in low temperature PL spectra (excitation 325 nm) indicate a higher degree of unintentional impurity incorporation for growth on O-polar substrate.

To further study the influence of the ZnO substrate polarity, MB327 was grown on O-polar ZnO substrate, but with a high-quality ZnO buffer layer. This buffer layer has a very low defect density and a n-type carrier concentration below 10^{16} cm^{-3}. No significant inhomogeneities were resolved with 50x optical microscopy on the surface of the ZnO:N top layer. Figure 6.8b shows the Raman spectra taken on the ZnO substrate, on the ZnO buffer layer, and on the ZnO:N top layer, respectively. Within the experimental error of the applied Raman scattering method, no differences are observable in these spectra. The narrow peak width of the E_2(high) mode at about 437 cm^{-1}, the weak intensity of the A_1(LO) mode at about 577 cm^{-1}, and the weak background confirm the high structural quality of the layers. However, Raman spectroscopy indicates no nitrogen incorporation. In particular, no additional modes occur in the spectrum of the ZnO:N top layer. These results confirm that, under the used growth conditions, the O-polarity of the ZnO substrate hinders successful nitrogen incorporation, despite the high-quality ZnO buffer layer.

Chapter 7

Summary

(i) Pure ZnO: bulk crystals, disorder effects, and nanoparticles (Chapter 4)

Bulk ZnO single crystals were characterized by Raman spectroscopy in different scattering configurations. The ZnO phonon modes obeyed the corresponding symmetry selection rules and their peak widths and vibrational frequencies confirmed a high structural quality. A multi-phonon process observed in these Raman spectra was identified as the phonon difference E_2(high)-E_2(low) by its vibrational frequency and temperature-dependent intensity behavior. In comparison to the high-quality ZnO single crystals, polycrystalline ZnO of minor structural quality was analyzed. The corresponding Raman spectra revealed mixed orientations and a broad disorder band in the LO mode region.

The Raman resonance in ZnO was studied using bulk single crystals. First effects were already visible when tuning the exciting laser wavelength from green (514.5 nm) to blue (457.9 nm). Much stronger resonance was observed at excitation with 363.8 nm close to the band gap of 3.4 eV. The ZnO A_1(LO) phonon and the corresponding multiple phonon processes 2xA_1(LO), 3xA_1(LO), etc., dominated the Raman spectrum due to the strong resonance of Fröhlich scattering.

In order to study extrinsic disorder effects, ZnO was irradiated with argon ions. Again, the disorder signature of ZnO was identified as a broad band in the LO mode region, which scaled with the argon fluence. Furthermore, no broad mode was observed in the Raman spectra of ZnO irradiated with low argon fluence, but only a slightly intensified A_1(LO) mode occurred. From these findings and from symmetry considerations, the broad ZnO disorder band was attributed to an intensified A_1(LO) mode due to extrinsic Fröhlich scattering. Its broadening was explained by disorder-induced Raman scattering contributions from outside the Brillouin zone center. Thermal annealing was applied to analyze the healing effect on the irradiation-induced disorder. It was found that most crystal damage in these argon-irradiated ZnO samples was healed already by annealing at 500 °C.

ZnO nanoparticles were grown according to a newly developed synthesis procedure. In their Raman spectra, a quasi-LO phonon mode was observed with a vibrational frequency corresponding to a random orientation of the crystallites. In addition to the ZnO phonon modes, molecular vibrations of the organic stabilizer were detected in all Raman spectra of ZnO nanoparticles analyzed for this thesis. In order to obtain ZnO nanoparticles capped with functional ligands, the synthesis was modified by the addition of the dye molecule oracet blue. During the Raman experiments on these nanoparticles, laser-induced local heating led to thermal annealing effects, improving the crystal quality of the particles, but presumably harming the organic ligands simultaneously. The phonon peak shifts observed during these experiments were attributed to local heating effects. In particular, no optical phonon confinement was observed.

To further analyze a possible optical phonon confinement, a ZnO nanoparticle series was studied with different average diameters ranging from 2.0 nm to 16 nm. Despite literature reports of phonon confinement effects for larger ZnO particles, the phonons appeared at their bulk positions in the Raman spectra of particle ensembles with average diameters down to 3.2 nm. The pentanetrione-capped nanoparticles with an average diameter of only about 2.0 nm are among the smallest ZnO nanoparticles reported until today. No phonon mode could be detected for these nanoparticles, presumably due to the reduced structural quality of the nanocrystal cores and the optical properties of the surrounding organics.

(ii) Transition-metal-alloyed ZnO (Chapter 5)

Manganese-, cobalt-, iron-, vanadium-, and nickel-alloyed ZnO was studied by Raman spectroscopy and complementary methods. The samples were fabricated by ion implantation, vapor phase transport, molecular beam epitaxy, and other growth techniques. The experiments on cobalt-, iron-, and nickel-implanted ZnO were the first reported Raman studies on such systems. In the focus of the experiments were disorder effects, an analysis of the substitutional transition metal incorporation into ZnO, and the magnetic impact of secondary phase formation. Such secondary phases were observed in several systems, depending on the transition metal species, transition metal concentration, growth method, and after-growth treatment.

The effect of transition metal implantation on ZnO crystals was studied using high implantation doses of vanadium, manganese, iron, cobalt, and nickel. The broad $A_1(LO)$ disorder band was observed in the Raman spectra again, but much more intense than for the argon-irradiated ZnO. Its intensity was especially strong for manganese implantation, which was attributed to reduced transparency of the manganese-implanted layer compared to the other transition metals. Thermal annealing of up to 500 °C reduced the implantation-induced crystal damage, but did not yet lead to a complete healing.

For manganese concentrations $\leq$ 8 at.% in ZnO, two contributions were identified by electron paramagnetic resonance: a fine structure due to isolated, substitutional Mn^{2+} ions and a broadened, unstructured signal from dipole-interacting, substitutional Mn^{2+} ions in the implanted layer. The isolated ions were already observed in the unimplanted ZnO crystals and therefore attributed to residual impurities of the host crystals. By photoluminescence experiments, iron ions were identified as additional residual impurities. In the Raman spectra of the manganese-alloyed ZnO crystals, the $A_1(LO)$ disorder band scaled with the manganese concentration. Much stronger resonance effects of this band were found in this manganese-alloyed ZnO compared to pure ZnO, which was explained by impurity-induced Fröhlich scattering. For very low manganese concentrations, most of the resolved features in the Raman spectra could be assigned to ZnO multi-phonon processes. One feature was suggested as a localized vibration of substitutional manganese. This assignment is based on the intensity behavior of this feature upon increasing manganese concentration and thermal annealing. Furthermore, of all transition-metal-alloyed ZnO samples studied in this thesis, only the manganese-alloyed samples exhibited this feature. It was found that the broad $A_1(LO)$ disorder band is red-shifted for very high disorder, corresponding to the dispersion of the $A_1(LO)$ phonon branch near the Brillouin zone center in the phonon dispersion relation of ZnO. By a Raman scattering depth profile, the disorder in the manganese-implanted layer was analyzed in detail. No secondary phases were detected by Raman scattering, in accordance with the findings of high-resolution transmission electron microscopy and X-ray diffraction. An additional mode, which occurred in the low-temperature Raman spectra, was identified as a ZnO-related, localized defect mode.

For implanted manganese concentrations $\geq$16 at.%, secondary phase formation was clearly observed after annealing at 700 °C in air. In the samples with the highest manganese concentration of 32 at.% relative to zinc, secondary phase islands of μm size were littered over the entire surface. As major secondary phase, $ZnMn_2O_4$ was identified by Raman scattering, in accordance with X-ray diffraction results. On a few localized spots of the sample surface, micro-Raman mapping revealed non-stoichiometric $Zn_{3-x}Mn_xO_4$ precipitates with Raman signatures similar to pure Mn_3O_4. These inclusions lay below the sensitivity limit of X-ray diffraction. High-resolution transmission electron microscopy and energy-dispersive X-ray spectroscopy line scans confirmed the presence of sub-μm, manganese-rich secondary phases with elongated shape and in alignment with the ZnO wurtzite structure.

The wet-chemical synthesis of manganese-alloyed ZnO nanoparticles was found to be very sensitive to the choice of the organic stabilizer. Nanoparticles capped with the ligand molecule DMPDA showed the characteristic Raman signature of manganese-alloyed ZnO, but with very broad and red-shifted phonon features. By annealing experiments, this effect could be attributed to local heating, alloy potential fluctuations, and disorder. Magnetic characterization and electron paramagnetic resonance measurements indicated that part of the manganese ions were located on the intended zinc lattice site. However, clustering manganese ions were also observed by these methods, presumably within the surrounding organics.

ZnO implanted with cobalt concentrations ≤ 8 at.% showed the characteristic $A_1(LO)$ disorder band in Raman scattering. Additionally, photoluminescence features were observed, which were attributed to substitutional cobalt incorporation on zinc sites. In ZnO:Co samples grown by vapor phase transport, no secondary phases could be identified by room-temperature Raman experiments. In experiments at 10 K, on the other hand, additional features occurred, which correspond to magnon Raman scattering from CoO inclusions within the crystal. Furthermore, substitutionally incorporated Co^{2+} ions were identified by photoluminescence experiments. In accordance with these results, electronic Raman scattering features were observed for the first time, which correspond to intra-3D ground level splitting of Co^{2+} in CoO and in ZnO.

For ZnCoO layers grown by molecular beam epitaxy, substitutional cobalt incorporation was confirmed by photoluminescence experiments. These layers showed ferromagnetic properties for small cobalt concentrations. For higher cobalt concentrations, they lost their ferromagnetic ordering and, accordingly, the photoluminescence signature of substitutional Co^{2+} was reduced and the Raman signature of ZnO disorder increased.

For a very high implanted cobalt concentration of 32 at.%, additional Raman modes were observed. Their vibrational frequencies and intensity ratios were found to be identical with the Raman features obtained from experiments with bulk $ZnCo_2O_4$, which was grown for comparison. These $ZnCo_2O_4$ precipitates turned out to be below the sensitivity limit of X-ray diffraction. Though, X-ray diffraction experiments revealed the presence of the secondary phases hcp zinc and hcp cobalt.

Nanocrystalline ZnCoO layers were grown by dip-coating and showed extremely strong crystal disorder. This disorder could be reduced by thorough thermal annealing. On the other hand, annealing at 700 °C induced the formation of $ZnCo_2O_4$ and Co_3O_4 secondary phases, whereas annealing at 900 °C led to the reduction of these cobalt oxides into CoO on the surface of the layers. Moreover, the nanocrystalline layers showed a very strong Raman resonance in the LO mode region due to impurity-induced Fröhlich scattering, while no resonance was observed for the secondary phase modes. With micro-Raman mapping, an inhomogeneous distribution of the formed cobalt oxides was found.

For iron-alloyed ZnO, secondary phase formation set in already for iron concentrations of 16 at.%, but again only after 700 °C annealing in air. High-resolution transmission electron microscopy revealed sub-μm precipitates of elongated shape, which were aligned to the ZnO wurtzite structure. X-ray diffraction measurements indicated that $ZnFe_2O_4$ was the major secondary phase in these samples. In contrast, no clear Raman signature was visible, but only broad and unstructured additional bands, which suggested the absence of well-ordered, stoichiometric iron oxides. For the highest iron concentration of 32 at.%, very strong disorder effects occurred after the 700 °C annealing in air, including even a partial surface peel-off. In order to evaluate the impact of the environment during the annealing procedure, additional experiments were conducted with iron-alloyed ZnO samples annealed in vacuum instead of air. This procedure resulted in a strongly improved surface quality and less pronounced secondary phase segregation. However, the formation of iron-related oxides was still not completely suppressed for high iron concentrations.

Nickel-alloyed ZnO showed the characteristic $A_1(LO)$ disorder Raman band scaling with the nickel concentration. At the highest nickel concentration of 32 at.% and after 700 °C annealing, secondary phase formation of NiO and cubic elemental nickel was identified by X-ray diffraction. Both phases cannot be observed by Raman scattering due to selection rules.

It was concluded that secondary phase formation in transition-metal-alloyed ZnO can be induced by thermal annealing, but can also occur during growth. Most of the identified secondary phases had a large impact on the magnetic properties of the entire system, even if they appeared only as sporadic, localized inclusions. Conventional X-ray diffraction was not sensitive enough to detect such singular inclusions, in contrast to micro-Raman mapping. On the other hand, some elemental phases and cubic transition metal oxides have no Raman signature. Therefore, a combination of these methods was required for the analysis of secondary phase formation in transition-metal-alloyed ZnO.

(iii) Nitrogen-doped ZnO (Chapter 6)

ZnO single crystals were implanted with nitrogen ions and characterized by Raman scattering and photoluminescence. A strong donor-acceptor-pair transition band in the photoluminescence spectra indicated the successful incorporation of nitrogen ions as dopants. Five strong additional modes appeared in the Raman spectra of nitrogen-doped ZnO for nitrogen concentrations >1 at.% relative to O. Their origin is disputed in literature and was therefore discussed in detail in this thesis. Structural disorder caused by the ion implantation was successfully healed by thermal annealing in vacuum at 700 °C and 900 °C. No intensity reduction of the nitrogen-related additional modes occurred upon this after-growth treatment. Furthermore, they were found to scale with the nitrogen concentration after the last annealing step. By a recently reported isotopic study, localized vibrations of substitutional nitrogen were excluded as the origin of these modes. Comparison to data reported in the literature clearly showed that the observed features do not correspond to the vibrations of N_2 or NO on oxygen sites and neither to nitrogen-hydrogen complexes. The favored explanation in literature is a pure disorder effect. However, this was disproved by the annealing experiments of this thesis, which did not lead to a substantial reduction of the additional modes. Therefore, and in accordance with theoretical considerations, the additional modes were attributed to impurity-activated B_1 silent modes of ZnO and the corresponding multi-phonon processes. The activation mechanism was found to be characteristic for nitrogen since the additional modes were not present in the Raman spectra of any of the ZnO systems alloyed with other elements than nitrogen in this thesis. Raman experiments in different scattering configurations were conducted to confirm the B_1(high) mode assignment of the additional mode with vibrational frequency close to the A_1(LO) phonon. In these experiments, the additional mode was observed for scattering configurations, in which the A_1(LO) mode is forbidden. Additionally, the symmetry selection rule for the A_1(LO) mode was confirmed by Raman experiments on a manganese-alloyed sample. They showed a strong A_1(LO) band in the allowed, but no A_1(LO) signal at all in the symmetry-forbidden scattering configuration.

Nitrogen-doped ZnO grown by molecular beam epitaxy was studied by Raman spectroscopy and complementary methods. Heteroepitaxial growth on sapphire led to strong disorder on the sample surface and micro-Raman mapping revealed an inhomogeneous distribution of the nitrogen-related additional modes. Homoepitaxial growth, in contrast, resulted in strongly improved structural properties. A large difference was observed between the homoepitaxial growth on the Zn-polar and O-polar surfaces of the ZnO substrate. While no additional modes were observed for growth on the O-polar surface, they appeared very strong if the Zn-polar surface was chosen. Photoluminescence and transport experiments showed that the growth on Zn-polar surfaces led to semi-insulating ZnO by the compensation of the ZnO-intrinsic donor levels. In contrast, no doping effects could be observed for growth on the O-polar surface, even if a high-quality ZnO buffer layer was grown between the substrate and the nitrogen-doped layer. The presence of the nitrogen-activated silent ZnO B_1 modes in Raman scattering was identified as a potential indicator for substitutional nitrogen incorporation.

Chapter 8

Zusammenfassung

(i) Reines ZnO: Volumenkristalle, Unordnungseffekte und Nanopartikel (Kapitel 4)

Zur Analyse ihrer Gitterdynamik-Eigenschaften wurden ZnO-Volumenkristalle mittels Ramanspektroskopie in verschiedenen Streukonfigurationen charakterisiert. Die ZnO-Phononmoden befolgten dabei die Symmetrie-Auswahlregeln von idealem ZnO und die beobachteten Peakbreiten und Schwingungsfrequenzen bestätigten eine hohe strukturelle Qualität der Proben. In den Ramanspektren dieser Einkristalle wurde ein Multiphonon-Prozess beobachtet, welcher aufgrund seiner Schwingungsfrequenz und seines temperaturabhängigen Verhaltens als die Phonon-Differenz E_2(high)-E_2(low) identifiziert werden konnte. Als Vergleich zu den hochqualitativen ZnO-Einkristallen wurde polykristallines ZnO niedrigerer Qualität untersucht. Die entsprechenden Ramanspektren zeigten gemischte Orientierungen und eine breite Unordnungsbande im Bereich der LO-Moden.

Die Raman-Resonanz in ZnO wurde anhand von volumenartigen Einkristallen untersucht. Erste Effekte waren schon bei einer Änderung der Laserwellenlänge vom grünen (514,5 nm) zum blauen (457,9 nm) Spektralbereich sichtbar. Eine viel stärkere Resonanz trat bei der Anregung mit 363,8 nm auf, und damit nahe der ZnO-Bandkante von 3,4 eV. Das ZnO-A_1(LO)-Phonon und die entsprechenden Multiphonon-Prozesse 2xA_1(LO), 3xA_1(LO) usw. dominieren das Ramanspektrum aufgrund der starken Resonanz der Fröhlich-Streuung.

Um extrinsische Unordnungseffekte zu untersuchen, wurde ZnO mit Argon-Ionen bestrahlt. Dabei konnte wiederum die Unordnungs-Signatur von ZnO als eine breite Bande im Bereich der LO-Moden identifiziert werden, welche mit dem Argon-Fluss skalierte. Dagegen wurde keine breite Mode in den Raman-Spektren der ZnO-Kristalle beobachtet, welche mit niedrigem Argon-Fluss bestrahlt worden waren, sondern es trat nur eine leicht verstärkte A_1(LO)-Mode auf.

Mit diesen Ergebnissen und durch Symmetrie-Überlegungen konnte die breite ZnO-Unordnungs-Bande einer intensivierten $A_1(LO)$-Mode aufgrund extrinsischer Fröhlich-Streuung zugeordnet werden. Ihre Verbreiterung wurde mit Unordnungs-induzierter Ramanstreuung von außerhalb des Brillouin-Zonen-Zentrums erklärt. Im Anschluss an die Ar-Bestrahlung wurden die Proben bei verschiedenen Temperaturen thermisch ausgeheilt und der zugehörige Heilungseffekt analysiert. Dabei stellte sich heraus, dass die Beschädigungen des Kristalls in diesen Argon-bestrahlten ZnO-Proben schon bei 500 °C weitestgehend ausgeheilt werden konnten.

ZnO-Nanopartikel wurden auf Basis einer neu entwickelten Synthese-Prozedur gewachsen. In den entsprechenden Ramanspektren trat ein Quasi-LO-Phonon mit einer Schwingungsfrequenz auf, welche einer zufälligen Orientierung der Kristallite entspricht. Neben den ZnO-Phononmoden wurden in den Ramanspektren aller ZnO-Nanopartikel dieser Arbeit auch Molekülschwingungen der organischen Stabilisatoren detektiert. Um ZnO-Nanopartikel zu erhalten, welche mit einem funktionalen Liganden stabilisiert sind, wurde die Synthese durch die Zugabe des Farbstoff-moleküls Oracet-Blau modifiziert. Während der Raman-Experimente an diesen Nanopartikeln führte laserinduzierte, lokale Aufheizung zu thermischen Ausheileffekten. Diese verbesserten zwar einerseits die Kristallqualität der Partikel, beschädigten aber gleichzeitig auch die organischen Liganden. Die Verschiebung der Phonon-Peaks, welche während dieser Experimente beobachtet wurde, konnte vollständig durch lokale Aufheizungseffekte erklärt werden. Insbesondere wurde kein Confinement der optischen Phononen beobachtet.

Um ein mögliches Phonon-Confinement weitergehend zu untersuchen, wurde eine Serie von ZnO-Nanopartikeln mit durchschnittlichen Durchmessern von 2,0 nm bis 16 nm spektroskopiert. Für alle Partikel-Ensembles mit durchschnittlichen Durchmessern ≥3,2 nm erschienen die Phononen an den Frequenz-Positionen des entsprechenden Volumenmaterials. Dies widerspricht Literaturberichten von Phonon-Confinement-Effekten in größeren ZnO-Partikeln. Die Pentanetrion-stabilisierten Nanopartikel mit einem durchschnittlichen Durchmesser von nur ca. 2,0 nm gehören zu den kleinsten jemals hergestellten ZnO-Nanopartikeln. Allerdings konnten für diese Nanopartikel keine Phononen detektiert werden, vermutlich aufgrund reduzierter struktureller Qualität der Nanopartikel-Kerne und bedingt durch die optischen Eigenschaften der umgebenden Organik.

(ii) ZnO legiert mit Übergangsmetallen (Kapitel 5)

Mangan-, Cobalt-, Eisen-, Vanadium-, und Nickel-legiertes ZnO wurde mittels Ramanspektroskopie und komplementärer Methoden untersucht. Hergestellt wurden die entsprechenden Proben u.a. durch Ionenimplantation, Gasphasen-Transport und Molekularstrahl-Epitaxie. Die

Experimente an Cobalt-, Eisen-, und Nickel-implantiertem ZnO stellen die ersten veröffentlichten Ramanstudien an solchen Systemen dar. Im Mittelpunkt der Experimente standen Unordnungseffekte, der substitutionelle Übergangsmetall-Einbau sowie die magnetische Auswirkung potentieller Fremdphasenbildung. Solche Fremdphasen wurden in mehreren der untersuchten Systeme beobachtet, abhängig vom jeweiligen Übergangsmetall-Element, der Übergangsmetall-Konzentration, dem Wachstumsprozess und von der Behandlung nach dem Wachstum.

Der Effekt der Übergangsmetall-Implantation auf ZnO-Kristalle wurde mittels hoher Implantations-Dosen der verwendeten Elemente untersucht. Auch in diesen Ramanspektren wurde die breite $A_1(LO)$-Unordnungsbande beobachtet, allerdings deutlich stärker als für das Argonbestrahlte ZnO. Besonders deutlich trat sie im Fall der Mangan-Implantation hervor, was einer reduzierten Transparenz der Mangan-implantierten Schicht verglichen mit anderen Übergangsmetallen zugeordnet wurde. Thermisches Ausheilen bei bis zu 500 °C reduzierte die Implantationsinduzierten Kristallschäden, führte jedoch noch nicht zu einer kompletten Heilung.

Bei der Elektronenspinresonanz-Analyse von ZnO legiert mit Mangan-Konzentrationen ≤ 8 at.% (relativ zu Zink) wurden zwei Beiträge identifiziert: eine Feinstruktur aufgrund isolierter, substitutioneller Mn^{2+}-Ionen und ein verbreitertes, unstrukturiertes Signal von Dipol-wechselwirkenden, substitutionellen Mn^{2+}-Ionen in der implantierten Schicht. Die isolierten Ionen wurden bereits in den unimplantierten ZnO-Kristallen beobachtet und daher wachstumsbedingten Fremdatomen der Wirtskristalle zugeordnet. Mittels Photolumineszenz-Experimenten wurden darüber hinaus auch kleine Konzentrationen von Eisen-Ionen in den nominell reinen ZnO-Kristallen nachgewiesen. In den Ramanspektren der Mangan-legierten ZnO-Kristalle skalierte die $A_1(LO)$-Unordnungsbande deutlich mit der Mangan-Konzentration. Außerdem wurde für diese Bande in Manganlegiertem ZnO eine viel größere Ramanresonanz festgestellt als in reinem ZnO, was mit Fremdatom-induzierter Fröhlich-Streuung erklärt werden konnte. Bei sehr niedrigen Mangan-Konzentrationen konnten die meisten der aufgelösten Ramansignale ZnO-Multiphonon-Prozessen zugeordnet werden. Eines dieser Signale wurde allerdings als mögliche lokalisierte Schwingung substitutionellen Mangans identifiziert. Diese Zuweisung beruht auf dem Intensitätsverhalten dieses Signals bei steigender Mangankonzentration und bei thermischem Ausheilen. Des Weiteren trat dieses Signal unter allen untersuchten Übergangsmetall-legierten ZnO-Proben nur bei Mangan-Legierung auf. Es wurde herausgefunden, dass die breite $A_1(LO)$-Unordnungsbande bei sehr starker Unordnung rotverschoben ist. Diese Verschiebung entspricht der Dispersion des $A_1(LO)$-Phononzweiges nahe dem Brillouin-Zonen-Zentrum in der Phononendispersionsrelation von ZnO. Mit einem Raman-Tiefenprofil wurde die Unordnung in der Mangan-implantierten Schicht im Detail analysiert. Mittels Ramanstreuung konnten keine Fremdphasen nachgewiesen werden, was durch hochauflösende Transmissionselektronenmikroskopie und Röntgenbeugung bestätigt wurde. In den entsprechenden Tieftemperatur-Ramanspektren trat eine zusätzliche Mode auf, welche als lokalisierte ZnO-Defektmode identifiziert werden konnte.

Für Mangan-Konzentrationen ≥ 16 at.% wurde nach thermischer Ausheilung bei 700 °C an Luft eine deutliche Fremdphasenbildung beobachtet. In den Proben mit der höchsten Mangan-Konzentration von 32 at.% waren ca. μm-große Fremdphasen-Inseln über die gesamte Oberfläche verstreut. Als hauptsächliche Fremdphase wurde $ZnMn_2O_4$ mittels Ramanstreuung identifiziert und durch Röntgenbeugungs-Experimente bestätigt. An wenigen Stellen der Probenoberfläche zeigten sich bei Mikroraman-Scans nicht-stöchiometrische $Zn_{3-x}Mn_xO_4$-Ablagerungen mit einer Raman-Signatur ähnlich dem reinen Mn_3O_4. Diese Einschlüsse lagen unterhalb der Empfindlichkeitsgrenze der Röntgenbeugung. Hochauflösende Transmissionselektronenmikroskopie und Energie-dispersive Röntgenspektroskopie bestätigten das Vorhandensein sub-μm-großer, manganreicher Fremdphasen länglicher Gestalt und ausgerichtet an der ZnO-Wurtzitstruktur.

Die nasschemische Synthese von Mangan-legierten ZnO-Nanopartikeln stellte sich als sehr empfindlich gegenüber der Wahl des organischen Stabilisators heraus. Nanopartikel stabilisiert mit dem Liganden DMPDA zeigten die charakteristische Raman-Signatur von Mangan-legiertem ZnO, allerdings mit sehr breiten und rotverschobenen Phononmoden. In Temper-Experimenten konnte dieser Effekt mit lokaler Erwärmung und dem Fremdatomeinbau erklärt werden. Magnetische Charakterisierung und Elektronenspinresonanz-Messungen zeigten, dass ein Teil der Mangan-Ionen auf den beabsichtigten Zink-Gitterplätzen eingebaut waren. Allerdings wurden mit diesen Methoden auch Mangan-Cluster beobachtet, vermutlich innerhalb der umgebenden Organik.

ZnO implantiert mit Cobalt-Konzentrationen ≤ 8 at.% zeigte die charakteristische $A_1(LO)$-Unordnungsbande in der Ramanstreuung. Zusätzlich wurden Photolumineszenz-Signale beobachtet, welche einem substitutionellen Einbau von Cobalt auf Zink-Plätzen zugeordnet wurden. In ZnO:Co-Proben, welche mittels Gasphasen-Transport gewachsen wurden, konnten in Raumtemperatur-Ramanexperimenten keine Fremdphasen identifiziert werden. In Experimenten bei 10 K dagegen traten zusätzliche Signale auf, welche Magnon-Ramanstreuung von CoO-Einschlüssen im Kristall entsprechen. Außerdem wurden durch Photolumineszenz-Experimente substitutionell eingebaute Co^{2+}-Ionen identifiziert. Übereinstimmend mit diesen Ergebnissen wurde zum ersten Mal elektronische Ramanstreuung beobachtet, welche die intra-3D-Grundzustandsaufspaltung von Co^{2+} in CoO und ZnO widerspiegelt.

Auch in ZnCoO-Schichten, gewachsen mittels Molekularstrahlepitaxie, bestätigten Photolumineszenz-Experimente einen substitutionellen Cobalt-Einbau. Diese Schichten zeigten ferromagnetische Eigenschaften bei Cobalt-Konzentrationen ≤ 3 at.%. Bei höheren Cobalt-Konzentrationen verloren sie allerdings ihre ferromagnetische Ordnung. Im Einklang damit sank die Photolumineszenz-Signatur von substitutionell eingebautem Co^{2+} und die Ramansignatur für ZnO-Unordnung nahm zu.

Bei einer sehr hohen implantierten Cobalt-Konzentration von 32 at.% wurden zusätzliche Ramanmoden beobachtet. Die Schwingungsfrequenzen und Intensitätsverhältnisse stimmten dabei genau mit den Raman-Signalen von Volumen-$ZnCo_2O_4$ überein, welches zum Vergleich gewachsen wurde. Die damit identifizierten $ZnCo_2O_4$-Ablagerungen waren unterhalb der Empfindlichkeitsgrenze der Röntgenbeugung. Andererseits wiesen Röntgenbeugungsexperimente elementares Zink (hcp) und Cobalt (hcp) als weitere Fremdphasen nach.

Nanokristalline ZnCoO-Schichten wurden mittels Tauchbeschichtung hergestellt und zeigten extrem starke Kristallunordnung. Diese Unordnung konnte durch gründliches thermisches Ausheilen reduziert werden. Allerdings induzierte das Tempern bei 700 °C die Bildung der Fremdphasen $ZnCo_2O_4$ und Co_3O_4 und die Behandlung bei 900 °C führte zur Reduktion dieser Cobaltoxide zu CoO auf der Probenoberfläche. Die nanokristallinen Schichten zeigten eine sehr starke Ramanresonanz im Bereich der LO-Moden aufgrund von Fremdatom-induzierter Fröhlich-Streuung. Für die Fremdphasen-Moden wurde dagegen keine Resonanz beobachtet. Mikroraman-Scans ergaben eine stark inhomogene Verteilung der gebildeten Cobaltoxide.

Im Fall von Eisen-legiertem ZnO setzte die Fremdphasenbildung schon bei Eisen-Konzentrationen von 16 at.% ein, allerdings wiederum erst nach thermischem Ausheilen bei 700 °C an Luft. Hochauflösende Transmissionselektronenmikroskopie zeigte längliche, sub-μm-große Ablagerungen, welche an der ZnO-Wurtzit-Struktur ausgerichtet waren. Röntgenbeugungs-Messungen ergaben, dass $ZnFe_2O_4$ die vorrangige Fremdphase in diesen Proben darstellte. Im Gegensatz dazu war keine eindeutige Ramansignatur sichtbar, sondern nur eine breite und unstrukturierte zusätzliche Bande, was die Abwesenheit geordneter, stöchiometrischer Eisenoxide nahelegt. Für die höchste Eisenkonzentration von 32 at.% traten nach dem Ausheilprozess bei 700 °C an Luft sehr starke Unordnungseffekte auf. Es wurde sogar eine teilweise Ablösung der Probenoberfläche beobachtet. Um den Einfluss der Temper-Umgebung zu analysieren, wurden zusätzliche Experimente mit Eisen-legiertem ZnO nach Tempern im Vakuum anstatt an Luft durchgeführt. Durch diesen Ausheilvorgang verbesserte sich die Oberflächenqualität deutlich und auch die Fremdphasenbildung war weniger stark ausgeprägt. Trotzdem wurde die Bildung oxidischer Fremdphasen bei hohen Eisen-Konzentrationen nicht gänzlich unterdrückt.

Bei der höchsten Nickel-Konzentration von 32 at.% und nach thermischem Ausheilen bei 700 °C wurde mittels Röntgenbeugung die Fremdphasenbildung von NiO und kubischem, elementarem Nickel identifiziert. Beide Phasen können aufgrund von Auswahlregeln nicht in der Ramanstreuung beobachtet werden.

Als Schlussfolgerung ergab sich, dass Fremdphasenbildung in Übergangsmetall-legiertem ZnO durch thermisches Ausheilen hervorgerufen wird, je nach System aber auch schon beim Wachstum auftreten kann. Die meisten der identifizierten Fremdphasen können die magnetischen Eigenschaften des gesamten Systems stark beeinflussen, selbst wenn sie nur als vereinzelte, lokale Einschlüsse auftreten. Im Gegensatz zu Mikroraman-Scans war konventionelle Röntgenbeugung nicht empfindlich genug, solche singulären Einschlüsse zu detektieren. Andererseits besitzen einige elementare Fremdphasen und kubische Übergangsmetall-Oxide keine Ramansignatur. Daher stellte sich die Kombination dieser Methoden für die Untersuchung der Fremdphasenbildung in Übergangsmetall-legiertem ZnO als nötig heraus.

(iii) Stickstoff-dotiertes ZnO (Kapitel 6)

ZnO-Einkristalle wurden mit Stickstoff implantiert und mittels Ramanstreuung- und Photolumineszenz-Experimenten charakterisiert. Deutliche Donator-Akzeptor-Paar-Übergänge in den Photolumineszenz-Spektren zeigten einen erfolgreichen Einbau von Stickstoff-Ionen als Dotanden an. Die strukturelle Unordnung der Proben, verursacht von der Ionenimplantation, wurde mit thermischem Ausheilen im Vakuum erfolgreich behandelt. Bei Stickstoff-Konzentrationen ≥ 1 at.% relativ zu O traten fünf starke, zusätzliche Moden in den Ramanspektren des Stickstoffdotierten ZnO auf. Die Herkunft dieser Moden ist in der Fachliteratur umstritten und wurde daher in dieser Arbeit detailliert analysiert. Das Ausheilen bei bis zu 900 °C bewirkte keinen Intensitätsrückgang dieser Stickstoff-induzierten, zusätzlichen Moden. Des Weiteren skalierten sie nach dieser Behandlung deutlich mit der Stickstoffkonzentration. Anhand einer kürzlich veröffentlichten Isotopenstudie wurden lokalisierte Schwingungen substitutionellen Stickstoffs als Ursache für die zusätzlichen Moden ausgeschlossen. Durch einen Vergleich mit Literatur-Ergebnissen wurde gezeigt, dass die in dieser Arbeit beobachteten Signale nicht mit den Schwingungen von N_2 oder NO auf Sauerstoff-Plätzen übereinstimmen und auch nicht Stickstoff-Wasserstoff-Komplexen zugeordnet werden können. Die in der Fachliteratur favorisierte Erklärung ist ein reiner Unordnungseffekt. Allerdings wurde dies mit den Temper-Versuchen dieser Arbeit widerlegt, die keine klare Reduktion der zusätzlichen Moden bewirkten. Daher wurden die zusätzlichen Moden in Übereinstimmung mit theoretischen Überlegungen identifiziert als Fremdatom-aktivierte, eigentlich stille ZnO-B_1-Moden und entsprechende Multiphonon-Prozesse. Der Aktivierungsmechanismus ist dabei charakteristisch für Stickstoff, da die zusätzlichen Moden von allen legierten ZnO-Systemen dieser Arbeit ausschließlich in den Stickstoff-legierten auftraten. Raman-Experimente in verschiedenen Streu-Konfigurationen wurden durchgeführt, um die Zuweisung einer der zusätzlichen Moden als B_1(high) zu bestätigen, welche eine Schwingungsfrequenz sehr nahe am A_1(LO)-Phonon besitzt. In diesen Experimenten wurde die zusätzliche Mode in Streugeometrien beobachtet, in welchen die A_1(LO)-Mode verboten ist. Des Weiteren wurde die Auswahlregel der A_1(LO)-Mode bei Raman-Experimenten an Mangan-legiertem ZnO bestätigt. Diese zeigten eine starke A_1(LO)-Bande in der erlaubten, aber überhaupt kein A_1(LO)-Signal in der verbotenen Streugeometrie.

Stickstoffdotiertes ZnO wurde außerdem auch mittels Molekularstrahlepitaxie hergestellt und mit Ramanspektroskopie und komplementären Methoden untersucht. Heteroepitaktisches Wachstum auf Saphir führte zu einer starken Unordnung auf den Probenoberflächen, und Mikroraman-Scans zeigten eine inhomogene Verteilung der Stickstoff-induzierten zusätzlichen Moden. Im Gegensatz dazu wurden bei homoepitaktischem Wachstum deutlich verbesserte strukturelle Eigenschaften erreicht. Ein großer Unterschied zeigte sich zwischen dem homoepitaktischen Wachstum auf Zn-polarer und auf O-polarer Oberfläche des ZnO-Substrats. Während keine zusätzlichen Moden bei Wachstum auf der O-polaren Oberfläche beobachtet wurden, traten sie sehr stark auf, wenn die Zn-polare Oberfläche gewählt wurde. Photolumineszenz- und Transport-Experimente zeigten, dass das Wachstum auf Zn-polaren Oberflächen zu semi-isolierendem ZnO führte, also eine Kompensation der ZnO-intrinsischen Donatoren bewirkt wurde. Dagegen konnten keine Dotierungseffekte bei Wachstum auf O-polarer Oberfläche beobachtet werden, selbst wenn eine hochqualitative ZnO-Pufferschicht zwischen dem Substrat und der Stickstoff-dotierten Schicht gewachsen wurde. Die Anwesenheit der Stickstoff-aktivierten ZnO-B_1-Moden in der Ramanstreuung wurde als potentieller Indikator für substitutionellen Stickstoff-Einbau identifiziert.

Appendix A

Abbreviations

AM - Additional mode

APF - Alloy potential fluctuations

BZ - Brillouin zone

CB - Conduction band

CCD - Charge-coupled device

CW - Continuous wave

DACH - Diaminocyclohexane

DAP - Donor-acceptor pair

DFT - Density functional theory

DLS - Dynamic light scattering

DMS - Diluted magnetic semiconductor

DMPDA - Dimethylpropylenediamine

DOS - Density of states

EDX - Energy-dispersive X-ray spectroscopy

EPR - Electron paramagnetic resonance

FM - Ferromagnet(ism)

FT - Fourier transform

FWHM - Full width at half maximum

HCP - Hexagonal close-packed

HRTEM - High-resolution transmission electron microscopy

LM - Local mode

LO - Longitudinal optical

LPDOS - Local phonon density of states

LVM - Localized vibrational mode

MOKE - Magneto-optical Kerr effect

MREI - Modified random-element isodisplacement model

NIR - Near infrared

PDOS - Phonon density of states

PL - Photoluminescence

PDR - Phonon dispersion relation

RT - Room temperature

TEM - Transmission electron microscopy

TMAH - Tetramethylammonium hydroxide

TM - Transition metal

TO - Transverse optical

UV - Ultraviolet

VB - Valence band

XAFS - X-ray absorption fine structure spectroscopy

XAS - X-ray absorption spectroscopy

XPS - X-ray photoelectron spectroscopy

XRD - X-ray diffraction

Bibliography

[**Abraham 2005**]: D. W. Abraham, M. M. Frank, and S. Guha, *Appl. Phys. Lett.* **87**, 252502 (2005)

[**Alaria 2005**]: J. Alaria, P. Turek, M. Bernard, M. Bouloudenine, A. Berbadj, N. Brihi, G. Schmerber, S. Colis, and A. Dinia, *Chem. Phys. Lett.* **415**, 337Ű41 (2005)

[**Alaria 2006**]: J. Alaria, M. Bouloudenine, G. Schmerber, S. Colis, A. Dinia, P. Turek, and M. Bernard, *J. Appl. Phys.* **99**, 08M118 (2006)

[**Alim 2005/1**]: K. A. Alim, V. A. Fonoberov, and A. A. Balandin, *Appl. Phys. Lett.* **86**, 053103 (2005)

[**Alim 2005/2**]: K. A. Alim, V. A. Fonoberov, and M. Shamsa, *J. Appl. Phys.* **97**, 124313 (2005)

[**Anastassakis 1991**]: E. Anastassakis in *Diluted Magnetic Semiconductors*, edited by M. Jain, World Scientific, Singapore, 225 (1991)

[**Arguello 1969**]: C. A. Arguello, D. L. Rousseau, and S. P. S. Porto, *Phys. Rev.* **181**, 1351 (1969)

[**Artus 2007**]: L. Artus, R. Cusco, E. Alarcon-Llado, G. Gonzalez-Diaz, I. Martil, J. Jimenez, B. Wang, and M. Callahan, *Appl. Phys. Lett.* **90**, 181911 (2007)

[**Ashrafi 2000**]: A. B. M. A. Ashrafi, A. Ueta, A. Avramescu, H. Kumano, I. Suemune, Y.-W. Ok, and T.-Y. Seong, *Appl. Phys. Lett.* **76**, 550 (2000)

[**Bakin 2006**]: A. Bakin, A. El-Shaer, A. Che Mofor, M. Kreye, A. Waag, F. Bertram, J. Christen, M. Heuken, and J. Stoimenos, *J. Crys. Grow.* **287**, 7 (2006)

[**Bandyopadhyay 2004**]: S. Bandyopadhyay and M. Cahay, *Appl. Phys. Lett.* **85**, 1433 (2004)

[**Bass 1960**]: F. G. Bass and M. I. Kaganov, *Sovi. Phys. JETP* **10**, 986 (1960)

[**Bates 1962**]: C. H. Bates, W. B. White, and R. Roy, *Science* **137**, 993 (1962)

[**Behera 2008**]: D. Beheraa and B. S. Acharya, *J. Lum.* **128**, 1577 (2008)

[**Bergman 1999**]: L. Bergman, M. Dutta, C. Balkas, R. F. Davis, J. A. Christman, D. Alexson, and R. J. Nemanichc, *J. Appl. Phys.* **85**, 3535 (1999)

[**Bergman 2005**]: L. Bergman, X.-B. Chen, J. Huso, J. L. Morrison, and H. Hoeck, *J. Appl. Phys.* **98**, 093507 (2005)

[**Bhatti 2007**]: K. P. Bhatti, S. Chaudhary, D. K. Pandya, and S. C. Kashyap, *J. Appl. Phys.* **101**, 033902 (2007)

[**Billas 1994**]: I. M. L. Billas, A. Chatelain, and W. A. de Heer, *Science* **265**, 1682 (1994)

[**Blasco 2006**]: J. Blasco and J. Garcia, *J. Sol. Stat. Chem.* **179**, 2199 (2006)

[**Borse 1999**]: P. H. Borse, D. Srinivas, R. F. Shinde, S. K. Date, W. Vogel, and S. K. Kulkarni, *Phys. Rev. B* **60**, 8659Ü64 (1999)

[**Brüesch 1986**]: P. Brüesch, *Phonons: Theory and Experiments II, Experiments and Interpretation of Experimental Results*, Springer, Berlin (1986)

[**Buchholz 2005**]: D. B. Buchholz, R. P. H. Chang, J. H. Song, and J. B. Ketterson, *Appl. Phys. Lett.* **87**, 082504 (2005)

[**Buciuman 1999**]: F. Buciuman, F. Patcas, R. Craciun, and D. R. T. Zahn, *Phys. Chem. Chem. Phys.* **1**, 185 (1999)

[Bundesmann 2003]: C. Bundesmann, N. Ashkenov, M. Schubert, D. Spemann, T. Butz, E.M. Kaidashev, M. Lorenz, and M. Grundmann, *Appl. Phys. Lett.* **83**, 1974Ü6 (2003)

[Bundesmann 2005]: C. Bundesmann, *Phonons and plasmons in ZnO-based alloy and doped ZnO thin films studied by infrared spectroscopic ellipsometry and Raman scattering spectroscopy*, PhD thesis, University of Leipzig (2005)

[Calleja 1977]: J. M. Calleja and M. Cardona, *Phys. Rev. B* **16**, 3753 (1977)

[Cardona 2007]: *Light scattering in solids I-X (Topics in Applied Physics)*, edited by M. Cardona et al., Springer, Berlin (1983-2007)

[Chang 1968]: I. F. Chang and S. S. Mitra, *Phys. Rev.* **172**, 924 (1968)

[Chelikowsky 1986]: J. R. Chelikowsky and J. K. Burdett, *Phys. Rev. Lett.* **56**, 961 (1986)

[Chen 2005]: Z. Q. Chen, A. Kawasuso, Y. Xu, H. Naramoto, X. L. Yuan, T. Sekiguchi, R. Suzuki, and T. Ohdaira, *J. Appl. Phys.* **97**, 013528 (2005)

[Chen 2006]: Z. Q. Chen, M. Maekawa, A. Kawasuso, S. Sakai, and H. Naramoto, *J. Appl. Phys.* **99**, 093507 (2006)

[Choopun 2007]: S. Choopun, N. Hongsith, E. Wongrat, T. Kamwanna, S. Singkarat, P. Mangkorntong, N. Mangkorntong, and T. Chairuangsri, *J. Am. Ceram. Soc.* **91**, 174 (2007)

[Chory 2007]: C. Chory, R. B. Neder, V. I. Korsunskiy, F. Niederdraenk, C. Kumpf, E. Umbach, M. Schumm, M. Lentze, J. Geurts, G. Astakhov, W. Ossau, and G. Müller, *phys. stat. sol c* **4**, 3260 (2007)

[Chou 1976]: H.-H. Chou and H. Y. Fan, *Phys. Rev. B.* **13**, 3924 (1976)

[Chourpa 2005]: I. Chourpa, L. Douziech-Eyrolles, L. Ngaboni-Okassa, J.-F. Fouquenet, S. Cohen-Jonathan, M. Souce, H. Marchais, and P. Dubois, *Analyst* **130**, 1395 (2005)

[Coey 2005]: J. M. Coey, M. Venkatesan, and C. B. Fitzgerald, *Nat. Mater.* **4**, 173 (2005)

[Colwell 1972]: P. J. Colwell and M. V. Klein, *Phys. Rev. B* **6**, 498 (1972)

[**Cong 2006**]: C. J. Cong, L. Liao, Q. Y. Liu, J. C. Li, and K. L. Zhang, *Nanotech.* **17**, 1520Ű6 (2006)

[**Cullity 1978**]: B. D. Cullity, *Elements of X-Ray Diffraction*, 2nd edition, Addison-Wesley, Reading (1978)

[**Cusco 2007**]: R. Cusco, E. Alarcon-Llado, J. Ibanez, L. Artus, J. Jimenez, B. Wang, and M. J. Callahan, *Phys. Rev. B* **75**, 165202 (2007)

[**Dalpian 2006**]: G. M. Dalpian and J. R. Chelikowsky, *Phys. Rev. Lett.* **96**, 226802 (2006)

[**Damen 1966**]: T. C. Damen, S. P. S. Porto, and B. Tell, *Phys. Rev.* **142**, 570 (1966)

[**De Faria 1997**]: D. L. A. De Faria, S. Venancio Silva, and M. T. de Oliveira, *J. Raman Spectrosc.* **28**, 873 (1997)

[**Demtröder 2002**]: W. Demtröder, *Laser spectroscopy - Basic Concepts and Instrumentation*, 2nd edition, Springer, Berlin (2002)

[**Dietl 2000**]: T. Dietl, H. Ohno, F. Matsukura, J. Cibert, and D. Ferrand, *Science 287*, 1019Ű22 (2000)

[**Dietz 1971**]: R. E. Dietz, G. I. Parisot, and A. E. Meixner, *Phys. Rev. B* **4**, 2302 (1971)

[**Du 2005**]: G. Du, Y. Ma, Y. Zhang, and T. Wang, *Appl. Phys. Rev.* **87**, 213103 (2005)

[**Du 2006**]: C. L. Du, Z. B. Gu, M. H. Lu, J. Wang, S. T. Zhang, J. Zhao, G. X. Cheng, H. Heng, and Y. F. Chen, *J. Appl. Phys.* **99**, 123515 (2006)

[**Dürr 2008**]: J. Dürr, *Charakterisierung von Akzeptoren in Zinkoxid - Characterization of acceptors in zinc oxide*, diploma thesis, University of Göttingen (2008)

[**Elliot 1963**]: R. J. Elliot and R. Loudon, *Phys. Rev. Lett.* **3**, 189 (1963)

[**Englman 1966**]: R. Englman and R. Ruppin, *Phys. Rev. Lett.* **16**, 898 (1966)

[**Erhart 2006**]: P. Erhart and K. Albe, *Appl. Phys. Lett.* **88**, 201918 (2006)

[**Esser 1996**]: N. Esser and J. Geurts in *Optical Characterization of Epitaxial Semiconductor Layers*, edited by G. Bauer and W. Richter, Springer, Berlin, 129 (1996)

[**Exarhos 1995**]: G. J. Exarhos and S. K. Sharma, *Thin Sol. Films* **270**, 27 (1995)

[**Ferrari 2006**]: A. C. Ferrari, J. C. Meyer, V. Scardaci, C. Casiraghi, M. Lazzeri, F. Mauri, S. Piscanec, D. Jiang, K. S. Novoselov, S. Roth, and A. K. Geim, *Phys. Rev. Lett.* **97**, 187401 (2006)

[**Fleury 1968**]: P. A. Fleury and R. Loudon, *Phys. Rev.* **166**, 514 (1968)

[**Fonoberov 2004**]: V. A. Fonoberov and A. A. Balandin, *Phys. Rev. B* **70**, 233205 (2004)

[**Friedrich 2007**]: F. Friedrich and N. H. Nickel, *Appl. Phys. Lett.* **91**, 111903 (2007)

[**Furdyna 1988**]: J. K. Furdyna, *J. Appl. Phys.* **64**, R29Ű64 (1988)

[**Gallant 2006**]: D. Gallant, M. Pezolet, and S. Simard, *J. Phys. Chem. B* **110**, 6871 (2006)

[**Garcia 2005**]: M. A. Garcia, M. L. Ruiz-Gonzalez, A. Quesada, J. L. Costa-Krämer, J. F. Fernandez, S. J. Khatib, A. Wennberg, A. C. Caballero, M. S. Martín-Gonzalez, M. Villegas, F. Briones, J. M. Gonzalez-Calbet, and A. Hernando, *Phys. Rev. Lett.* **94**, 217206 (2005)

[**Gebicki 2005**]: W. Gebicki, K. Osuch, C. Jastrzebski, Z. Golacki, and M. Godlewski, *Superlatt. and Microstruct.* **38**, 428Ű38 (2005)

[**Geurts 1996**]: J. Geurts, *Prog. Crys. Grow. and Charact.* **32**, 185 (1996)

[**Haboeck 2005**]: U. Haboeck, A. Hoffmann, C. Thomsen, A. Zeuner, and B. K. Meyer, *phys. stat. sol. b* **242**, R21Ű3 (2005)

[**Hadjiev 1988**]: V. G. Hadjiev, M. N. Iliev, and I. V. Vergilov, *J. Phys. C* **21**, L199 (1988)

[**Hayes 1979**]: W. Hayes, R. Loudon, and J. F. Scott, *Am. J. Phys.* **47**, 571 (1979)

[**Heitz 1992**]: R. Heitz, A. Hoffmann, and I. Broser, *Phys. Rev. B* **45**, 8977 (1992)

[**Hewat 1970**]: A.W. Hewat, *Sol. Stat. Comm.* **8**, 187 (1970)

[**Hong 2007**]: N. H. Hong, J. Sakai, and V. Brize, *J. Phys.: Cond. Mat.* **19**, 036219 (2007)

[**Ip 2003**]: K. Ip, R. M. Frazier, Y. W. Heo, D. P. Norton, C. R. Abernathy, S. J. Pearton, J. Kelly, R. Rairigh, A. F. Hebard, J. M. Zavada, and R. G. Wilson, *J. Vac. Sci. Technol. B* **21**, 1476 (2003)

[**Jagadish 2006**]: *Zinc Oxide Bulk, Thin Films and Nanostructures: Processing, Properties and Applications*, 1st edition, edited by C. Jagadish and S. J. Pearton, Elsevier, Oxford (2006)

[**Jawhari 1995**]: T. Jawhari, A. Roid, and J. Casado, Carbon 33, 1561 (1995)

[**Jeong 2004**]: T. S. Jeong, M. S. Han, C. J. Youn, and Y. S. Park, *J. Appl. Phys.* **96**, (2004)

[**JCPDS 1997**]: Joint Committee for Powder Diffraction Studies (JCPDS), ICDD, PDF2 Database PCPDFWIN 1.30 (1997)

[**Jin 2000**]: Z. Jin, M. Murakami, T. Fukumura, Y. Matsumoto, A. Ohtomo, M. Kawasaki, and H. Koinuma, *J. Crys. Grow.* **214**, 55 (2000)

[**Jin 2001**]: Z. Jin, T. Fukumura, M. Kawasaki, K. Ando, H. Saito, T. Sekiguchi, Y. Z. Yoo, M. Murakami, Y. Matsumoto, T. Hasegawa, and H. Koinuma, *Appl. Phys. Lett.* **78**, 3824 (2001)

[**Jouanne 2006**]: M. Jouanne, J. F. Morhange, W. Szuszkiewicz, Z. Golacki, and A. Mycielski, *phys. stat. sol. c* **3**, 1205Ű8 (2006)

[**Julien 2004**]: C. M. Julien, M. Massot, and C. Poinsignon, *Spectrochim. Act. A* **60**, 689 (2004)

[**Karmakar 2007**]: D. Karmakar, S. K. Mandal, R. M. Kadam, P. L. Paulose, A. K. Rajarajan, T. K. Nath, A. K. Das, I. Dasgupta, and G. P. Das, *Phys. Rev. B* **75**, 144404 (2007)

[**Kaschner 2002**]: A. Kaschner, U. Haboeck, Mar. Strassburg, Mat. Strassburg, G. Kaczmarczyk, A. Hoffmann, C. Thomsen, A. Zeuner, H. R. Alves, D. M. Hofmann, and B. K. Meyer, *Appl. Phys. Lett.* **80**, 1909 (2002)

[**Kauschke 1987**]: W. Kauschke, A. K. Sood, M. Cardona, and K. Ploog, *Phys. Rev. B* **36**, 1612 (1987)

[**Kennedy 1995**]: T. A. Kennedy, E. R. Glaser, P. B. Klein, and R. N. Bhargava, *Phys. Rev. B.* **52**, R14356 (1995)

[**Kim 2004**]: H. J. Kim, I. C. Song, J. H. Sim, H. Kim, D. Kim, Y. E. Ihm, and W. K. Choo, *phys. stat. sol. b* **241**, 1553 (2004)

[**Kittilstved 2005**]: K. R. Kittilstved, N. S. Norberg, and D. R. Gamelin, *Phys. Rev. Lett.* **94**, 147209 (2005)

[**Klingshirn 2007**]: C. Klingshirn, *phys. stat. sol. b* **244**, 3027 (2007)

[**Knickelbein 2001**]: M. B. Knickelbein, *Phys. Rev. Lett.* **86**, 5255 (2001)

[**Kodama 1997**]: R. H. Kodama, S. A. Makhlouf, and A. E. Berkowitz, *Phys. Rev. Let.* **79**, 1393 (1997)

[**Kohan 2000**]: A. F. Kohan, G. Ceder, D. Morgan, and C. G. Van de Walle, *Phys. Rev. B* **61**, 15019 (2000)

[**Koidl 1977**]: P. Koidl, *Phys. Rev. B* **15**, 2493 (1977)

[**Kolesnik 2004**]: S. Kolesnik, B. Dabrowski, and J. Mais, *J. Appl. Phys.* **95**, 2582Ű6 (2004)

[**Kucheyev 2001**]: S. O. Kucheyev, J. S. Williams, and S. J. Pearton, *Mater. Sci. Eng. R* **33**, 51 (2001)

[**Kucheyev 2006**]: S. O. Kucheyev and C. Jagadish in *Zinc Oxide Bulk, Thin Films and Nanostructures: Processing, Properties and Applications*, 1st edition, edited by C. Jagadish and S. J. Pearton, Elsevier, Oxford, 285 (2006)

[**Kumpf 2005**]: C. Kumpf, R. B. Neder, F. Niederdraenk, P. Luczak, A. Stahl, M. Scheuermann, S. Joshi, S. K. Kulkarni, C. Barglik-Chory, C. Heske, and E. Umbach, *J. Chem. Phys.* **123**, 224707 (2005)

[**Kuzian 2006**]: R. O. Kuzian, A. M. Dare, P. Sati, and R. Hayn, *Phys. Rev. B* **74**, 155201 (2006)

[**Lakshmi 2006**]: P. V. B. Lakshmi and K. Ramachandran, *Radiat. Eff. Def. Sol.* **161**, 365Ű71 (2006)

[**Lautenschlaeger 2008**]: S. Lautenschlaeger, J. Sann, N. Volbers, B. K. Meyer, A. Hoffmann, U. Haboeck, and M. R. Wagner, *Phys. Rev. B* **77**, 144108 (2008)

[**Lawes 2005**]: G. Lawes, A. S. Risbud, A. P. Ramirez, and R. Seshadri, *Phys. Rev. B* **71**, 045201 (2005)

[**Lee 2001**]: E.-C. Lee, Y.-S. Kim, Y.-G. Jin, and K. J. Chang, *Phys. Rev. B* **64**, 085120 (2001)

[**Lee 2002**]: H.-J. Lee, S.-Y. Jeonga, C. R. Cho, and C. H. Park, *Appl. Phys. Lett.* **81**, 4020-2 (2002)

[**Limpijumnong 2005**]: S. Limpijumnong, X. Li, S.-H. Wei, and S. B. Zhang, *Appl. Phys. Lett.* **86**, 211910 (2005)

[**Lin 2004**]: C.-C. Lin, S.-Y. Chen, S.-Y. Cheng, and H.-Y. Lee, *Appl. Phys. Lett.* **84**, 5040 (2004)

[**Liu 2005**]: C. Liu, F. Yun, and H. Morkoc, *J. Mater. Sci.: Mater. Electron.* **16**, 555Ű97 (2005)

[**Look 2005**]: D. C. Look, G. C. Farlow, P. Reunchan, S. Limpijumnong, S. B. Zhang, and K. Nordlund, *Phys. Rev. Lett.* **95**, 225502 (2005)

[**Look 2006**]: D. C. Look in *Zinc Oxide Bulk, Thin Films and Nanostructures: Processing, Properties and Applications*, 1st edition, edited by C. Jagadish and S. J. Pearton, Elsevier, Oxford, 21 (2006)

[**Loudon 1964**]: R. Loudon, *Adv. in Phys.* **13**, 423 (1964)

[**Lu 2006**]: J. Lu, Q. Zhang, J. Wang, F. Saito, and M. Uchida, *Powd. Tech.* **162**, 33 (2006)

[**Manjon 2005**]: F. J. Manjon, B. Mari, J. Serrano, and A. H. Romero, *J. Appl. Phys.* **97**, 053516 (2005)

[**Mayur 1996**]: A. J. Mayur, M. D. Sciacca, H. Kim, I. Miotkowski, A. K. Ramdas, S. Rodriguez, and G. C. La Rocca, *Phys. Rev. B* **53**, 12884 (1996)

[**McCluskey 2000**]: M. D. McCluskey, *J. Appl. Phys.* **87**, 3593 (2000)

[**McCreery 2000**]: R. L. McCreery, *Raman Spectroscopy for Chemical Analysis*, 1st edition, Wiley interscience, New York (2000)

[**Meyer 2004**]: B. K. Meyer, H. Alves, D. M. Hofmann, W. Kriegseis, D. Forster, F. Bertram, J. Christen, A. Hoffman, M. Straßburg, M. Dworzak, U. Haboeck, and A. V. Rodina, *phys. stat. sol. b* **241**, 231Ű60 (2004)

[**Millot 2006**]: M. Millot, J. Gonzalez, I. Molina, B. Salas, Z. Golacki, J. M. Broto, H. Rakoto, and M. Goiran, *J. Alloy. and Comp.* **423**, 224 (2006)

[**Mofor 2005**]: A. C. Mofor, A. El-Shaer, A. Bakin, A. Waag, H. Ahlers, U. Siegner, S. Sievers, M. Albrecht, W. Schoch, N. Izyumskaya, and V. Avrutin, *Appl. Phys. Lett.* **87**, 062501 (2005)

[**Mofor 2006**]: A. C. Mofor, F. Reuss, A. El-Shaer, H. Ahlers, U. Siegner, A. Bakin, W. Limmer, J. Eisenmenger, Th. Mueller, P. Ziemann, and A. Waag, *phys. stat. sol. c* **3**, 1104 (2006)

[**Monteiro 2003**]: T. Monteiro, C. Boemare, M. J. Soares, E. Rita, and E. Alves, *J. Appl. Phys.* **93**, 8995 (2003)

[**Muck 2004**]: T. Muck, J. W. Wagner, L. Hansen, V. Wagner, J. Geurts, and S. V. Ivanov, *Phys. Rev. B* **69**, 245314 (2004)

[**Neder 2005**]: R. B. Neder and V. I. Korsunskiy, *J. Phys.: Cond. Mat.* **17**, 125Ű34 (2005)

[**Neder 2007**]: R. B. Neder, V. I. Korsunskiy, Ch. Chory, G. Müller, A. Hofmann, S. Dembski, Ch. Graf, and E. Rühl, *phys. stat. sol. c* **4**, 1Ü13 (2007)

[**Nickel 2003**]: N. H. Nickel and K. Fleischer, *Phys. Rev. Lett.* **90**, 197402 (2003)

[**Niederdraenk 2007**]: F. Niederdraenk, K. Seufert, P. Luczak, S. K. Kulkarni, C. Chory, R. B. Neder, and C. Kumpf, *phys. stat. sol. c* **4**, 3234Ü43 (2007)

[**Norton 2003**]: D. P. Norton, M. E. Overberg, S. J. Pearton, K. Pruessner, J. D. Budai, L. A. Boatner, M. F. Chisholm, J. S. Lee, Z. G. Khim, Y. D. Park, and R. G. Wilson, *Appl. Phys. Lett.* **83**, 5488 (2003)

[**Norton 2006**]: D. P. Norton, S. J. Pearton, J. M. Zavada, W. M. Chen, and I. A. Buyanova in *Zinc Oxide Bulk, Thin Films and Nanostructures: Processing, Properties and Applications*, 1st edition, edited by C. Jagadish and S. J. Pearton, Elsevier, Oxford, 555 (2006)

[**O'Neill 1985**]: H. St. C. O'Neill, *Phys. Chem. Min.* **12**, 149 (1985)

[**Ozgur 2005**]: U. Ozgur, Y. I. Alivov, C. Liu, A. Teke, M. A. Reshchikov, S. Dogan, V. Avrutin, S. J. Cho, and H. Morkoc, *J. Appl. Phys.* **98**, 041301 (2005)

[**Parayanthal 1984**]: P. Parayanthal and F. H. Pollak, *Phys. Rev. Lett.* **52**, 1822 (1984)

[**Park 2004**]: J. H. Park, M. G. Kim, H. M. Jang, S. Ryu, and Y. M. Kim, *Appl. Phys. Lett.* **84**, 1338 (2004)

[**Peterson 1986**]: D.L. Peterson, A. Petrou, W. Giriat, A.K. Ramdas, and S. Rodriguez, *Phys. Rev. B* **33**, 1160 (1986)

[**Petrou 1983**]: A. Petrou, D. L. Peterson, S. Venugopalan, R. R. Galazka, A. K. Ramdas, and S. Rodriguez, *Phys. Rev. B* **27**, 3471 (1983)

[**Pfeiffer 2007**]: N. Pfeiffer, *Synthese und Charakterisierung von ZnO-Nanopartikeln - Synthesis and characterization of ZnO nanoparticles*, diploma thesis, University of Würzburg (2007)

[**Phan 2007**]: T. L. Phan, R. Vincent, D. Cherns, N. X. Nghia, M. H. Phan, and S. C. Yu, *J. Appl. Phys.* **101**, 09H103 (2007)

[**Philipose 2006**]: U. Philipose, S. V. Naira, S. Trudel, C. F. de Souza, S. Aouba, R. H. Hill, and H. E. Ruda, *Appl. Phys. Lett.* **88**, 263101 (2006)

[**Phillips 1970**]: J. C. Phillips, *Rev. Mod. Phys.* **42**, 317 (1970)

[**Piekarczyk 1988**]: W. Piekarczyk, P. Peshev, A. Toshev, and A. Pajaczkowska, *Mat. Res. Bull.* **23**, 1299 (1988)

[**Polyakov 2003**]: A. Y. Polyakov, A. V. Govorkov, N. B. Smirnov, N. V. Pashkova, S. J. Pearton, M. E. Overberg, C. R. Abernathy, D. P. Norton, J. M. Zavada, and R. G. Wilson, *Sol. Stat. Electr.* **47**, 1523 (2003)

[**Potzger 2008**]: K. Potzger, K. Kuepper, Q. Xu, S. Zhou, H. Schmidt, M. Helm, and J. Fassbender, *J. Appl. Phys.* **104**, 023510 (2008)

[**Rajalakshmi 2000**]: M. Rajalakshmi, A. K. Arora, B. S. Bendre, and S. Mahamuni, *J. Appl. Phys.* **87**, 2445 (2000)

[**Ramdas 1982**]: A. K. Ramdas, *J. Appl. Phys.* **53**, 7649 (1982)

[**Rao 2005**]: C. N. R. Rao and F. L. Deepak, *J. Mater. Chem.* **15**, 573 (2005)

[**Raskin 2008**]: M. Raskin, *Mikroramanspektroskopie an ZnO-Nanopartikeln mit organischen Liganden - Micro-Raman spectroscopy on ZnO nanoparticles with organic ligands*, diploma thesis, University of Würzburg (2008)

[**Reeber 1970**]: R. R. Reeber, *J. Appl. Phys.* **41**, 5063 (1970)

[**Reuss 2004**]: F. Reuss, C. Kirchner, Th. Gruber, R. Kling, S. Maschek, W. Limmer, A. Waag, and P. Ziemann, *J. Appl. Phys.* **95**, 3385 (2004)

[**Richter 1976**]: W. Richter in *Springer Tracts in Modern Physics Vol. 78: Solid State Physics*, edited by G. Höhler, Springer, Berlin, 121 (1976)

[**Richter 1981**]: H. Richter, Z. P. Wang, and L. Ley, *Sol. Stat. Comm.* **39**, 625 (1981)

[**Rita 2004**]: E. Rita, U. Wahl, J. G. Correia, E. Alves, J. C. Soares, and The ISOLDE Collaboration, *Appl. Phys. Lett.* **85**, 4899 (2004)

[**Roth 1958**]: W. L. Roth, *Phys. Rev.* **110**, 1333 (1958)

[**Ruppin 1970**]: R. Ruppin and R. Englman, *Rep. Prog. Phys.* **33**, 149 (1970)

[**Saeki 2001**]: H. Saeki, H. Tabata, and T. Kawai, *Sol. Stat. Comm.* **120**, 439 (2001)

[**Sakurai 1968**]: J. Sakurai, W. J. L. Buyers, R. A. Cowley, and G. Dolling, *Phys. Rev.* **167**, 510Ű8 (1968)

[**Samanta 2006**]: K. Samanta, P. Bhattacharya, R. S. Katiyar, W. Iwamoto, P. G. Pagliuso, and C. Rettori, *Phys. Rev. B* **73**, 245213 (2006)

[**Samanta 2007**]: K. Samanta, S. Dussan, R. S. Katiyar, and P. Bhattacharya, *Appl. Phys. Lett.* **90**, 261903 (2007)

[**Sati 2006**]: P. Sati, R. Hayn, R. Kuzian, S. Regnier, S. Schäfer, A. Stepanov, C. Morhain, C. Deparis, M. Lauügt, M. Goiran, and Z. Golacki, *Phys. Rev. Lett.* **96**, 017203 (2006)

[**Sato 2001**]: K. Sato and H. Katayama-Yoshida, *Jpn. J. Appl. Phys.* **40**, L334 (2001)

[**Sato 2002**]: K. Sato and H. Katayama-Yoshida, *phys. stat. sol. b* **229**, 673Ű80 (2002)

[**Sato-Berru 2007**]: R. Y. Sato-Berru, A. Vazquez-Olmos, A. L. Fernandez-Osorio, and S. Sotres-Martinez, *J. Raman Spectrosc.* **38**, 1073Ű6 (2007)

[**Scherrer 1918**]: P. Scherrer, *Nachr. Ges. Wiss. Göttingen* **26**, 98 (1918)

[**Schneider 1962/1**]: J. Schneider and S. R. Sircar, *Z. Naturf. A* **17**, 570 (1962)

[**Schneider 1962/2**]: J. Schneider and S. R. Sircar, *Z. Naturf. A* **17**, 651 (1962)

[**Schumm 2005**]: M. Schumm, *Mikroramanspektroskopie an II/VI-Halbleiternanopartikeln - Micro-Raman spectroscopy on II/VI semiconductor nanoparticles*, diploma thesis, University of Würzburg (2005)

[**Schumm 2007**]: M. Schumm, M. Koerdel, J. F. Morhange, Z. Golacki, K. Grasza, P. Skupinski, W. Szuszkiewicz, H. Zhou, V. Malik, H. Kalt, C. Klingshirn, and J. Geurts, *J. Phys.: Conf. Ser.* **92**, 012149 (2007)

[**Schumm 2008/1**]: M. Schumm, M. Koerdel, S. Müller, H. Zutz, C. Ronning, J. Stehr, D. M. Hofmann, and J. Geurts, *New J. Phys.* **10**, 043004 (2008)

[**Schumm 2008/2**]: M. Schumm, M. Koerdel, S. Mueller, C. Ronning, E. Dynowska, Z. Golacki, W. Szuszkiewicz, and J. Geurts, *Secondary phase segregation in heavily transition metal implanted ZnO*, *J. Appl. Phys.* (submitted)

[**Scott 1969**]: J. F. Scott, R. C. C. Leite, and T. C. Damen, *Phys. Rev.* **188**, 1285 (1969)

[**Scott 1970**]: J. F. Scott, *Phys. Rev. B* **2**, 1209 (1970)

[**Serrano 2003**]: J. Serrano, F. Widulle, A. H. Romero, M. Cardona, R. Lauck, and A. Rubio, *phys. stat. sol. b* **235**, 260 (2003)

[**Serrano 2004**]: J. Serrano, A. H. Romero, F. J. Manjon, R. Lauck, M. Cardona, and A. Rubio, *Phys. Rev. B* **69**, 094306 (2004)

[**Serrano 2007**]: J. Serrano, F. J. Manjon, A. H. Romero, A. Ivanov, R. Lauck, M. Cardona, and M. Krisch, *phys. stat. sol. b* **244**, 1478Ű82 (2007)

[**Seshadri 2005**]: R. Seshadri, *Curr. Opin. Sol. Stat. Mater. Sci.* **9**, 1Ű7 (2005)

[**Sharma 2003**]: P. Sharma, A. Gupta, K.V. Rao, F. J. Owens, R. Sharma, R. Ahuja, J. M. O. Guillen, B. Johansson, and G. A. Gehring, *Nat. Mat.* **2**, 673 (2003)

[**Shi 2007**]: T. Shi, S. Zhu, Z. Sun, S. Wei, and W. Liu, *Appl. Phys. Lett.* **90**, 102108 (2007)

[**Shim 2005**]: J. H. Shim, T. Hwang, S. Lee, J. H. Park, S.-J. Han, and Y. H. Jeong, *Appl. Phys. Lett.* **86**, 082503 (2005)

[**Siegle 1997**]: H. Siegle, G. Kaczmarczyk, L. Filippidis, A. P. Litvinchuk, A. Hoffmann, and

C. Thomsen, *Phys. Rev. B* **55**, 7000 (1997)

[**Sievers 1988**]: A. J. Sievers and S. Takeno, *Phys. Rev. Lett.* **61**, 970 (1988)

[**Slater 1936**]: J. C. Slater, *Phys. Rev.* **49**, 537 (1936)

[**Sonder 1988**]: E. Sonder, R. A. Zuhr, and R. E. Valiga, *J. Appl. Phys.* **64**, 1140 (1988)

[**Spaldin 2004**]: N. A. Spaldin, *Phys. Rev. B* **69**, 125201 (2004)

[**Stein 1994**]: H. J. Stein, *Appl. Phys. Lett.* **64**, 1520 (1994)

[**Sudakar 2007**]: C. Sudakar, J. S. Thakur, G. Lawes, R. Naik, and V. M. Naik, *Phys. Rev. B* **75**, 054423 (2007)

[**Szuszkiewicz 2007**]: W. Szuszkiewicz, J. F. Morhange, Z. Golacki, A. Lusakowski, M. Schumm, and J. Geurts, *Act. Phys. Pol. A* **112**, 363 (2007)

[**Szuszkiewicz 2008**]: W. Szuszkiewicz, J.-F. Morhange, A. Lusakowski, Z. Golacki, M. Arciszewska, B. B. Brodowska, M. Klepka, and W. Dobrowolski, *Raman spectroscopy and magnetic properties of a bulk $Zn_{0.984}Co_{0.016}O$ crystal*, E-MRS Fall Meeting, Warsaw (2008)

[**Tamura 2003**]: K. Tamura, T. Makino, A. Tsukazaki, M. Sumiya, S. Fuke, T. Furumochi, M. Lippmaa, C. H. Chia, Y. Segawa, H. Koinuma, and M. Kawasaki, *Sol. Stat. Comm.* **127**, 265 (2003)

[**Thakur 2007**]: J. S. Thakur, G. W. Auner, V. M. Naik, C. Sudakar, P. Kharel, G. Lawes, R. Suryanarayanan, and R. Naik, *J. Appl. Phys.* **102**, 093904 (2007)

[**Thoma 1974**]: K. Thoma, B. Dorner, G. Duesing, and W. Wegener, *Sol. Stat. Comm.* **15**, 1111 (1974)

[**Thomas 1964**]: D. G. Thomas, M. Gershenzon, and F. A. Trumbore, *Phys. Rev.* **133**, A269 (1964)

[**Thurian 1995**]: P. Thurian, G. Kaczmarczyk, H. Siegle, R. Heitz, A. Hoffmann, I. Broser, B. K. Meyer, R. Hoffbauer, and U. Scherz, *Mater. Sci. Forum* **196Ű201**, 1571Ű6 (1995)

[**Tsai 2008**]: S.-Y. Tsai, Y.-M. Lu, and M.-H. Hon, *J. Phys.: Conf. Ser.* **100**, 042037 (2008)

[**Tsuboi 1964**]: M. Tsuboi, *J. Chem. Phys.* **40**, 1326 (1964)

[**Tsukazaki 2004**]: A. Tsukazaki, A. Ohtomo, T. Onuma, M. Ohtani, T. Makino, M. Sumiya, K. Ohtani, S. F. Chichibu, S. Fuke, Y. Segawa, H. Ohno, H. Koinuma, and M. Kawasaki, *Nature Materials* **4**, 42 (2004)

[**Tu 2006**]: M.-L. Tu, Y.-K. Su, and C.-Y. Ma, *J. Appl. Phys.* **100**, 053705 (2006)

[**Turrell 1996**]: G. Turrell, *Raman Microscopy and Applications*, 1st edition, edited by G. Turrell and J. Corset, Academic Press (1996)

[**Tzolov 2000**]: M. Tzolov, N. Tzenov, D. Dimova-Malinovska, M. Kalitzova, C. Pizzuto, G. Vitali, G. Zollo, and I. Ivanov, *Thin Sol. Films* **379**, 28 (2000)

[**Tzolov 2001**]: M. Tzolov, N. Tzenov, D. Dimova-Malinovska, M. Kalitzova, C. Pizzuto, G. Vitali, G. Zollo, and I. Ivanov, *Thin Sol. Films* **396**, 274 (2001)

[**Ueda 2001**]: K. Ueda, H. Tabata, and T. Kawai, *Appl. Phys. Lett.* **79**, 988 (2001)

[**Van de Walle 2000**]: C. G. Van de Walle, *Phys. Rev. Lett.* **85**, 1012 (2000)

[**Venkataraj 2007**]: S. Venkataraj, N. Ohashi, I. Sakaguchi, Y. Adachi, T. Ohgaki, H. Ryoken, and H. Haneda, *J. Appl. Phys.* **102**, 014905 (2007)

[**Venkatesan 2004**]: M. Venkatesan, C. B. Fitzgerald, J. G. Lunney, and J. M. D. Coey, *Phys. Rev. Lett.* **93**, 177206 (2004)

[**Wagner 2002**]: V. Wagner, J. Wagner, S. Gundel, L. Hansen, and J. Geurts, *Phys. Rev. Lett.* **89**, 166103 (2002)

[**Wang 2001**]: X. Wang, S. Yang, J. Wang, M. Li, X. Jiang, G. Du, X. Liu, and R. P. H. Chang, *J. Crys. Grow.* **222**, 123 (2001)

[**Wang 2004**]: Z. L. Wang, *J. Phys.: Cond. Mat.* **16**, R829Ű58 (2004)

[**Wang 2005**]: J. B. Wang, H. M. Zhong, Z. F. Li, and W. Lu, *J. Appl. Phys.* **97**, 086105 (2005)

[**Wang 2006/1**]: J. B. Wang, G. J. Huang, X. L. Zhong, L. Z. Sun, Y. C. Zhou, and E. H. Liu, *Appl. Phys. Lett.* **88**, 252502 (2006)

[**Wang 2006/2**]: J. B. Wang, H. M. Zhong, Z. F. Li, and W. Lu, *Appl. Phys. Lett.* **88**, 101913 (2006)

[**Wang 2007**]: X. Wang, J. Xu, X. Yu, K. Xue, J. Yu, and X. Zhao, *Appl. Phys. Lett.* **91**, 031908 (2007)

[**Wolf 2001**]: S. A. Wolf, D. D. Awschalom, R. A. Buhrman, J. M. Daughton, S. von Molnar, M. L. Roukes, A. Y. Chtchelkanova, and D. M. Treger, *Science* **294**, 1488 (2001)

[**Wolk 1993**]: J. A. Wolk, J. W. Ager III, K. J. Duxstad, E. E. Haller, N. R. Taskar, D. R. Dorman, and D. J. Olego, *Appl. Phys. Lett.* **63**, 2756 (1993)

[**Xu 2006**]: H. Y. Xu, Y. C. Liu, C. S. Xu, Y. X. Liu, C. L. Shao, and R. Mu, *J. Chem. Phys.* **124**, 074707 (2006)

[**Xu 2007**]: Q. Xu, H. Schmidt, L. Hartmann, H. Hochmuth, M. Lorenz, A. Setzer, P. Esquinazi, C. Meinecke, and M. Grundmann, *Appl. Phys. Lett.* **91**, 092503 (2007)

[**Yang 2005**]: L. W. Yang, X. L. Wu, G. S. Huang, T. Qiu, and Y. M. Yang, *J. Appl. Phys.* **97**, 014308 (2005)

[**Yu 1999**]: P. Y. Yu and M. Cardona, *Fundamentals of Semiconductors - Physics and Materials Properties*, 2nd edition, Springer, Berlin (1999)

[**Yu 2006**]: J. Yu, H. Xing, Q. Zhao, H. Mao, Y. Shen, J. Wang, Z. Lai, and Z. Zhu, *Sol. Stat. Comm.* **138**, 502 (2006)

[**Zhang 2006**]: Y. B. Zhang, S. Li, T. T. Tan, and H. S. Park, *Sol. Stat. Comm.* **137**, 142 (2006)

[**Zheng 2004**]: R. K. Zheng, H. Liu, X. X. Zhang, V. A. L. Roy, and A. B. Djurisic, *Appl.*

Phys. Lett. **85**, 2589 (2004)

[**Zhong 2006**]: H. Zhong, J. Wang, X. Chen, Z. Li, W. Xu, and W. Lu, *J. Appl. Phys.* **99**, 103905 (2006)

[**Zhou 2003**]: H. Zhou, D. M. Hofmann, A. Hofstaetter, and B. K. Meyer, *J. Appl. Phys.* **94**, 1965 (2003)

[**Zhou 2006**]: H. Zhou, D. M. Hofmann, H. R. Alves, and B. K. Meyer, *J. Appl. Phys.* **99**, 103502 (2006)

[**Zhou 2007/1**]: H. Zhou, L. Chen, V. Malik, C. Knies, D. M. Hofmann, K. P. Bhatti, S. Chaudhary, P. J. Klar, W. Heimbrodt, C. Klingshirn, and H. Kalt, *phys. stat. sol. a* **204**, 11 (2007)

[**Zhou 2007/2**]: S. Q. Zhou, K. Potzger, H. Reuther, G. Talut, F. Eichhorn, J. von Borany, W. Skorupa, M. Helm, and J. Fassbender, *J. Phys. D* **40**, 964 (2007)

[**Zhou 2008/1**]: S. Q. Zhou, K. Potzger, J. von Borany, R. Grötzschel, W. Skorupa, M. Helm, and J. Fassbender, *Phys. Rev. B* **77**, 035209 (2008)

[**Zhou 2008/2**]: S. Q. Zhou, K. Potzger, G. Talut, H. Reuther, J. von Borany, R. Grötzschel, W. Skorupa, M. Helm, J. Fassbender, N. Volbers, M. Lorenz, and T. Herrmannsdörfer, *J. Appl. Phys.* **103**, 023902 (2008)

[**Zhou 2008/3**]: S. Q. Zhou, K. Potzger, G. Talut, J. von Borany, W. Skorupa, M. Helm, and J. Fassbender, *J. Appl. Phys.* **103**, 07D530 (2008)

[**Ziegler 1985**]: J. F. Ziegler, J. P. Biersack, and U. Littmark, *The Stopping and Range of Ions in Solids*, Pergamon, New York (1985)

[**Zuo 2001**]: J. Zuo, C. Xu, L. Zhang, B. Xu, and R. Wu, *J. Raman Spectrosc.* **32**, 979 (2001)

www.ingramcontent.com/pod-product-compliance
Lightning Source LLC
Chambersburg PA
CBHW070714220326
41598CB00024BA/3147